AF718805

WITHDRAWN

Springer Series in Wood Science

Editor: T. E. Timell

Martin H. Zimmermann
Xylem Structure and the Ascent of Sap (1983)

John F. Siau
Transport Processes in Wood (1984)

Robert R. Archer
Growth Stresses and Strains in Trees (1986)

W. E. Hillis
Heartwood and Tree Exudates (1987)

Lorin W. Roberts/Peter B. Gahan/Roni Aloni
Vascular Differentiation and Plant Growth Regulators (1988)

Sherwin Carlquist
Comparative Wood Anatomy (1988)

L.W. Roberts P.B. Gahan R. Aloni

Vascular Differentiation and Plant Growth Regulators

With a Foreword by R.D. Preston

With 28 Figures

Springer series in wood science
cs/t

Springer-Verlag
Berlin Heidelberg New York London Paris Tokyo

K
25
R723
988

Prof. Dr. LORIN W. ROBERTS
University of Idaho, Department of Biological Sciences, Moscow, Idaho 83843, USA

Prof. Dr. PETER B. GAHAN
King's College London, Biology Department, Campden Hill Road, London W8 7AH, United Kingdom

Prof. Dr. RONI ALONI
Tel Aviv University, The George S. Wise Faculty of Life Sciences, Department of Botany, Tel Aviv, 69978, Israel

Series Editor:
T. E. TIMELL
State University of New York, College of Environmental, Science and Forestry, Syracuse, NY 13210, USA

Cover: Transverse section of *Pinus lambertiana* wood. Courtesy of Dr. Carl de Zeeuw, SUNY College of Environmental Science and Forestry, Syracuse, New York

ISBN 3-540-18989-0 Springer-Verlag Berlin Heidelberg New York
ISBN 0-387-18989-0 Springer-Verlag New York Berlin Heidelberg

Library of Congress Cataloging-in-Publication Data.

Roberts, Lorin Watson. Vascular differentiation and plant growth regulators. (Springer series in wood science) Bibliography; p.109. Includes index.
1. Vascular system of plants. 2. Plant cell differentiation. 3. Plant regulators. 4. Plant hormones. I. Gahan, Peter B. II. Aloni, R. (Roni), 1944– . III. Title. IV. Series.
QK725.R723 1988 582'.031 88-6661

This work is subject to copyright. All rights are reserved, whether the whole or part of the material is concerned, specifically the rights of translation, reprinting, re-use of illustrations, recitation, broadcasting, reproduction on microfilms or in other ways, and storage in data banks. Duplication of this publication or parts thereof is only permitted under the provisions of the German Copyright Law of September 9, 1965, in its version of June 24, 1985, and a copyright fee must always be paid. Violations fall under the prosecution act of the German Copyright Law.

© Springer-Verlag Berlin Heidelberg 1988
Printed in Germany

The use of registered names, trademarks, etc. in this publication does not imply, even in the absence of a specific statement, that such names are exempt from the relevant protective laws and regulations and therefore free for general use.

Typesetting, printing and bookbinding: Brühlsche Universitätsdruckerei, Giessen
2131/3130-543210

Foreword

The vascular tissue of higher plants has been an object of continuous detailed study since the invention of the light microscope. As relevant new physical, chemical and biochemical techniques and concepts have appeared over the years, they have regularly and immediately been applied to this field of enquiry. The reasons are not far to seek. Vascular tissue provides the long-distance transport system for water and nutrients within the plant, and an understanding of the structure and development of the pathways within it is essential for an understanding of growth in all higher plants, including crop plants. Moreover, parts of the vascular tissue – the whole of the xylem and the fibers of the problem – have been, and still are, in high demand commercially for their unique properties and as the only renewable structural material on earth. There are, in addition, more subtle ways in which xylogenesis, particularly as it allows the development of tree species, contributes to the environment. A few years ago, while walking in the foothills of the Pyrenees, I found all this expressed more elegantly than I can express it, in a plea carved in wood at the entrance to a forest:

Homme!!

Je suis la chaleur de ton foyer par les froids nuits
d'hiver, l'ombrage ami lorsque brûle le soleil d'été,
l'eau des rivières et des sources.
Je suis la charpente de ta maison, la planche de ta table.
Je suis le lit dans lequel tu dors et le bois dont tu
fais tes navires.
Je suis la manche de ta houe et la porte de ton enclos.
Je suis le bois de ton berceau et de ton cercueil.
Je suis le pain de ta bonté, la fleur de ta beauté,
Ecoute ma prière:
Ne me détruis pas, aime-moi.

This book shows a devotion to xylem development at a level that the author of this supplication cannot have conceived.

Xylogenesis presents an example of cell differentiation in an exceptionally complex form. It is concerned with the formation of primary and secondary vascular tissues as a whole, which conform in general to a plan common to all higher plants. At the same time, these tissues vary in topography between, for instance, angiosperms and gymnosperms, between monocotyledons and dicotyledons, and between angiosperms with different phyllotaxis; and in the details of their cellular composition to the extent that, in trees at least, species can be identified solely by the xylem. The process is, moreover, controlled by a wide variety of factors both exogenous and endogenous, including hormones, and by interactions between them. Investigations range, therefore, from studies of cell wall biosynthesis, through consideration of the fine tuning which induces neighboring cells to take different courses, to the control mechanisms which keep the whole in order. Some of these can, and must be examined with material in culture; some, on the other hand, can be considered only by the use of whole plants. There is, therefore, in addition to these a wide variety of experimentation involved, anatomical, physiological and biochemical, under a wide variety of conditions.

In such a situation it is inevitable that relevant observations are reported over a widely dispersed literature and the authors of this book have performed a valuable service in collecting these findings, including their own, by no means the least important, under one cover.

The hormone concept as applied to plants has a long history. The idea that one part of a plant may influence another part through the agency of a chemical messenger was first mooted by Sachs in 1880. It was not, however, until about 60 years ago that Went, inspired by the work of Boysen-Jensen, discovered the hormone auxin, and much more recently still that the effect of this hormone on xylogenesis was revealed. Since then, other hormones have been identified and this book deals with all of them.

The founding fathers of developmental anatomy, including such figures as Strasburger and Haberlandt, would have been entranced to see in this book concepts given reality which they could formulate only in the vaguest terms; and would not have been surprised that, in spite of all the work that has been done since their time, much remains still to be uncovered. My old mentor, the late Professor J. H. Priestley, himself an eminent developmental anatomist of his day who, to his dying day, refused to accept the existence of plant hormones (so that we members of staff were precluded from mentioning them in our teaching) would, at last I think, have been convinced.

Leeds, 1988 R. D. Preston

Preface

The purpose of this volume in the Springer Series in Wood Science is to present the highly complex nature of the regulation of vascular differentiation by growth regulators. The subject is presented by three "vascular scientists." Each of us has addressed the topic from a somewhat different viewpoint. Any study of hormone regulation in the whole plant is made problematic by the fact that every cell within the organism has its own unique polarity and, whether plastic or rigid in its state of differentiation, will have been acted upon by one or more plant growth regulators on one or more occasions. Thus, we are painters working on a canvas which has already been used by other artists. It is within this limitation that we try to identify bona fide hormonal cellular responses.

Because of the paucity of information on sieve element formation, the bulk of the book deals with the initiation of cytodifferentiation and subsequent development of tracheary elements. The opening chapter examines problems associated with the earliest programming of procambial cells in embryonic tissues, plus a myriad of complex issues on cytodifferentiation. One interesting topic is the employment of carboxylesterase activity as a biochemical marker acting as a reference for timing the stages of cell programming. The second chapter concerns the possible roles of the major groups of plant hormones during cytodifferentiation. The major emphasis is on the enigmatic role of auxin, and the recent evidence implicating calcium as an important second messenger. Another chapter is devoted to the organized differentiation of phloem and xylem within the whole plant in relation to vascular development and function. Evidence in favor of the Aloni and Zimmermann (1983) hypothesis is presented, as well as an outline of Aloni's (1987b) vascular adaptation hypothesis. Experimental data analyzed from studies on the induction of vascular differentiation in wounds and in cultured tissues are still another topic. The *Zinnia* system, in particular, has given us some important information on the biochemistry of xylogenesis. The final chapter involves the influences of physical factors on vascular differentiation. Since mechanical stress is associated with ethylene biosynthesis, the evidence supporting the view that ethylene is a hormone regulating tracheary element differentiation is presented in this chapter.

We are well aware of the limitations of a monograph on vascular differentiation, due to the lack of knowledge about the functional mechanics of auxin action. We have yet to devise a testable theory for the hormonal regulation of xylogenesis, although we can be reasonably sure that some of the pieces of the puzzle include auxin, cytokinin, ethylene, and calcium. Perhaps this book will provide the stimulus for other investigators to devise such a theory. A theory may not be forthcoming until we have developed the "ultimate in vitro system," that is, the induction of a tracheary element or a sieve element from a single isolated cell under rigidly controlled microenvironmental conditions and in the presence of precisely monitored hormonal levels. Nevertheless, we have tried to draw attention to the gaps in our knowledge, experimental systems available for further exploitation, and the apparent conflicts in our current knowledge. We hope that readers will be sufficiently stimulated to want to enter the fray and to reexamine a number of the existing problems.

Acknowledgments. The authors are indebted to several colleagues who have assisted us in the preparation of these chapters: Peter Barlow, Atsushi Komamine, Philip Larson, Raymond Miller, and Tsvi Sachs. We are particularly grateful for the guidance and encouragement provided by the Editor of this Series, Tore Timell. Financial assistance was received (PBG) by the Science and Engineering Research Council (U.K.), the Natural Environmental Research Council (U.K.), the Fonds National Suisse, and the Laboratoire de Physiologie Végétale, University of Genève. Finally, we appreciate the warm interest that Professor Preston has shown in our joint undertaking.

Spring, 1988

L. W. Roberts
P. B. Gahan
R. Aloni

Contents

Chapter 1

Xylem and Phloem Differentiation in Perspective

P. B. GAHAN

1.1 Introduction

The discovery of the microscope made possible an examination of the internal structure of multicellular plants, and the first publication on vascular plant anatomy involved the xylem tissue of the walnut tree (Henshaw 1661, cited in Preston 1981). Early publications on the tissues of higher plants by Grew (1682) and Malpighi (1675) also described mainly the xylem, since xylem forms the major supporting tissue of all the spermatophytes and pteridophytes. In spite of innumerable publications since those early days, Preston (1981) 300 years later commented that "more needs to be known about the differentiation of wood, anatomically, physically, and biochemically ...

Within the plant, vascular differentiation occurs in a variety of sites along its main axis and appendages (leaves and flowers). Many publications have reviewed the anatomical and cytological events occurring during the differentiation of xylem and phloem (von Sachs 1882, Bower 1930, Esau 1943, 1969, Eames and MacDaniels 1947, Johansen 1950, Cutter 1971, 1978, Foster and Gifford 1974). All of these studies have one important characteristic in common. There appears to be a pattern of vascular tissue differentiation, both primary and secondary, that is common to all members of the spermatophytes, gymnosperms, pteridophytes, and bryophytes. Once the vascular system is initiated only a simple series of subsequent steps need be added to maintain that system. Presumably we are seeking an operating pattern which is fundamental to all plants with vascular systems. Although detailed studies of the control of vascular differentiation have been limited to a relatively small number of species, there is a strong probability that the mechanisms described by such studies (for example, Sachs 1981, Aloni 1987 b) will be essentially common to all species.

Very little work has been performed on the initiation of the procambium within the embryo. These are cells which have been programmed for vascular tissue, and yet do not express all of this program until after seed germination. Most studies have concerned the main axis of the embryo after germination and the whole plant and its appendages. This is due, in part, to a lack of biochemical markers of procambial activity in embryonic tissues (Shininger 1979 b). Because of the economic concerns of agriculture and forestry, and difficulty in observing the phloem (Chap. 3.1.2), the emphasis of studies on whole plants has been on xylem formation. The major events accompanying the formation of primary (derived from primary meristems) and secondary (derived from cambia) vascular tissues, and the possible roles of plant growth regulators in these events, have been postulated by several workers (Fig. 1.1, Table 1.1). The possible cytological locations of biochemical events regulating cell wall metabolism are given (Fig. 1.2, Table 1.2), as well as possible control points in the cell cycle of pre- and early-

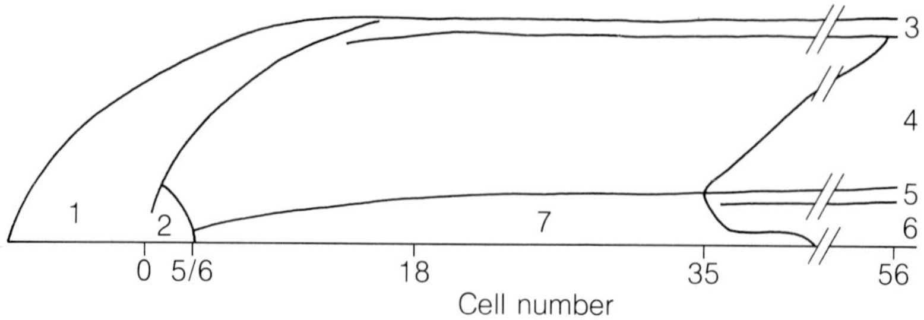

Fig. 1.1. Diagram of a longitudinally bisected root apex from a dicotyledonous plant showing the relationships between tissues, cells determined for vascular tissue, and specific physiological zones. Root cap (1) quiescent center (*2*) epidermis (*3*) cortex (*4*) endodermis (*5*) vascular tissues (*6*) carboxylesterase (*7*). Cell *0* is the junction of the root and root cap. Cells *5–6* are located immediately outside of the quiescent center and are determined to form vascular tissue. Cells *18–22* indicate the zone at which the capacity to produce secondary walls can be expressed. Cells *35–56* mark the zone of endoreduplication. Endoreduplication is common in angiosperms and occurs in algae and bryophytes, but is absent from gymnosperms. Phloem elements normally appear before xylem elements. (After publications by Tschermack-Woess 1956, Gahan 1981, McGarry and Gahan 1985)

Table 1.1. Postulated hormonal controls and biochemical events occurring during the ontogenetic stages of vascular element differentiation[a]

Stage of differentiation	Hormonal activity	Biochemical events
A. Competence of target cell. Cells determined early in embryogenesis Determined occurs at a given point in the cell cycle Primary meristems contain determined cells Secondary meristem cells are determined	Auxin and cytokinin play roles in cell cycle control and in gene expression. Sequential activity of these hormones can be postulated	Similar levels of glycolysis in all cells, but low levels of pentose phosphate pathway activity. High levels of polyamine formation from S-adenosylmethionine Hydrolyase activities limited. No genetic program available for biosynthesis of secondary wall Protein and nuclei acid synthesis
B. Cell enlargement. Cells exhibit variation in the extent and regulation of expansion. Deposition of primary wall material Assembly and positioning of microtubules for control of secondary wall pattern. Increase in the number of Golgi bodies	Auxin or a GA: cytokinin balance may be involved Activation of β-1,4-glucan glycosyl transferase by auxin Microtubule orientation may be regulated by gibberellin Gibberellin may induce hydrolyase activity and thus release monomers for wall biosynthesis. Ethylene may influence wall thickening and stimulate an increase in hydroxyproline-rich peroxidases in the developing cell wall	Synthesis of protein, lipid, nucleic acids, and nucleotide diphosphate sugars of the D-galactose series. Glycolysis as in A, but increased pentose phosphate pathway activity, UDP-glucose dehydrogenase activity, and production of hemicellulose precursors. Decrease in polyamine production, thus freeing S-adenosylmethionine for ethylene biosynthesis

Table 1 (continued)

Stage of differentiation	Hormonal activity	Biochemical events
C. Endoreduplication	Cytokinin may regulate this stage, or it may be regulated by the lack of this hormone	Biosynthesis of protein, nucleic acids, and phospholipids. High pentose phosphate pathway activity. Glycolysis rate similar to A
D. Secondary wall formation and lignification. Cell enlargement ceases Transfer of secondary wall components by Golgi-derived vesicles. Microtubules play role in orientation, but not in synthesis and deposition of cellulose microfibrils Controls set for hydrolase activation and cell autolysis Loss of nucleus from xylem and phloem elements. Phloem loses Golgi bodies, ribosomes, and microtubules	GA_3-induced hydrolyase activity may provide sugar monomers for cell walls. Auxin may regulate hemicellulose production. Possible roles for auxin, cytokinins, ethylene, and gibberellins in regulation of lignification. Sucrose may operate as a carbon source rather than as a hormone	Synthesis of secondary wall monomers with decreased epimerase activity. Synthesis of cellulose and hemicellulose. Activation of pathways for lignin biosynthesis. High velocity of pentose phosphate pathway for production of NADPH and shikimic acid. High PAL activity. Capacity to produce and express lignin-associated peroxidases
E. Autolysis of cell contents and dissolution of non-lignified parts of the cell wall. Degradation enzymes possibly released from lysosomes in the vacuoles. Breakdown of cytoplasmic membranes and organelles. Cytoplasmic debris lost from xylem elements	Auxin and cytokinin may be transported to adjacent cells together with other cytoplasmic material released by hydrolysis and having xylogenic activity	

[a] The contents of this table are modifications of previous views expressed by Torrey (1953), Roberts (1976), Northcote (1985a), and Gahan (1981).

differentiating cells (Table 1.3). A major problem in the interpretation of the roles of plant growth regulators in vascular differentiation is that in vitro studies permit us only to relate specific regulators with specific events. Each regulator can be involved in more than one event, and more than one regulator appears to be involved in any given event. Although there have been numerous studies on regulators acting as primary messengers, and other molecules (for example, calcium and polyamines) as secondary messengers, additional information is needed on binding sites and on the mechanisms of action of both types of messengers (Venis 1985).

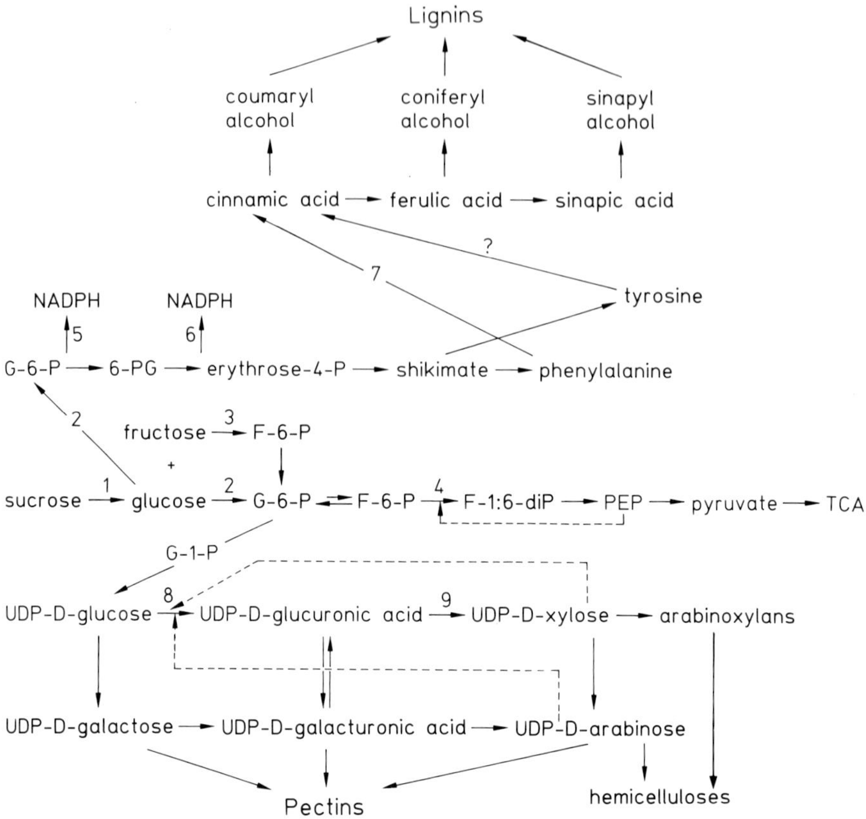

Fig. 1.2. Major reactions in the pathways leading to the production of pectins, hemicelluloses and lignin involved in cell wall formation. The major control steps are numbered. *Dotted lines* indicate feed-back mechanisms. Invertase (*1*), glucokinase (*2*), fructokinase (*3*), phosphofructokinase/phosphofructokinase: PP_i: fructose-2:6-bisphosphate (*4*), glucose-6-phosphate dehydrogenase (*5*), 6-phosphogluconate dehydrogenase (*6*), phenylalanine ammonia lyase (*7*), UDP-D-glucose dehydrogenase (*8*), UDP-D-glucuronic acid decarboxylase (*9*). (After Northcote 1982, 1984, 1985, McGarry and Gahan 1985, Gahan unpublished data)

Table 1.2. Possible cytological locations of biochemical events associated with the formation of primary and secondary cell walls

Location	Biochemical event
Cytosol	Nucleoside diphosphate sugars as precursors of polysaccharides
Golgi apparatus	Transglycolation for hemicelluloses and pectins; transport of nucleoside diphosphate sugars (?); transport of carboxylesterases
Endoplasmic reticulum	Epimerases for the interconversion of sugar nucleotides; initial glycosylations; transport of nucleoside diphosphate sugars (?); transport of acetylesterases
Plasmalemma	Synthases; spinning out of cellulose microfibrils; malate dehydrogenase
Cell wall	Malate dehydrogenase; peroxidases; synthases; cellulose microfibrils; carboxylesterases

Table 1.3. Cell cycle events and controls relevant to cytodifferentiation. (After Bryant 1976)

G_Q	G_1	S	G_2	M
May be present or absent	Late major control point at boundary with S phase May be present or absent Responsive to nonspecific S phase factors in the cell Decrease in level of poly (ADP-ribose) in chromatin Increase in accessibility of DNA to ribonuclease *Cells may differentiate*	Presence of specific and nonspecific factors promoting DNA synthesis. DNA and histone synthesis Very low poly (ADP-ribose) polymerase activity Increased NAD phosphorylase Decrease in accessibility of DNA to endonuclease Activity of newly synthesized DNA copies expressed Block in cycle leading to over replication of part of genome (e.g., protocorm of *Cymbidium*) Formation of polytene chromosomes at end of S phase (e.g., *Phaseolus*) Between S phase and metaphase the centromeres fail to divide and yield diplochromosomes (e.g., protocorm of *Cymbidium*) and differentiated cells in roots of higher plants Onset of chromatin synthesis at G_1-S transition	Presence of factors influencing onset of mitosis in early G_2 Decreased NAD phosphorylase activity in nucleus Increased levels of poly (ADP-ribose) in chromatin Activity of newly formed DNA copies expressed *Cells may differentiate*	No nuclear division results in polyploidy No cytokinesis yields binucleate cells Onset of chromatin condensation at G_2-M transition

1.2 Procambium Initiation in Embryos

Although many hypotheses have been advanced concerning the differentiation events, there is not as yet a definitive description of the control of xylem and phloem formation at the cellular and molecular levels. The primary concern has been with the maintenance of an existing and ongoing developmental process in the main axis of juvenile and adult plants (Larson 1983), rather than an examination of the initiation of cytodifferentiation within the embryonic axis. Primary xylem and phloem are produced from primary meristem cells. Secondary xylem and phloem arise from the vascular cambium, and this process is a continuation of a basic system already established by the procambium in the embryo. Information gleaned from studies on intact plants, wound responses, cultured plant material, and pathological conditions all have one characteristic in common, that is, the plant material was invariably derived from tissues in which the patterns of differentiation already existed. The observed response was preconditioned by the system employed. For example, vascular elements can be seen developing from preexisting prostelar tissue in cultured explants of the root apices of *Pisum sativum* (Rana and Gahan 1982). In order to understand the programming of xylem and phloem differentiation, it would be advantageous not only to examine procambial initiation during embryogenesis, but also to study the influences of the maternal tissue in this process. At this time the genetic program for vascular differentiation is set. This developmental process can be studied during embryo formation in situ or during somatic embryogenesis. The embryoids of the latter system, however, may be influenced in their development by the tissues from which they were derived. They may mature into plantlets, produce either roots or shoots, or fail to develop (G. P. Chapman et al. 1985).

The literature contains many examples of cell lineage analyses during embryo development, after fertilization of the ovum, for a broad range of plants (Johansen 1950, Johri 1984, Raghavan 1986). A problem common to all of these studies is the precise identification of the true procambium at its earliest inception (Shininger 1979 b). Although some workers have described the plerome histogen, which comprises the precursor cells of the procambium (Reeve 1948), this zone was identified by cytological changes and not by a biochemical marker. A biochemical marker enables the investigator to identify gene expression occurring prior to any visible cytological changes. Many workers have described the sequence of development as follows: stelar promeristem (plerome) → procambial initials → procambium (Bower 1930, Esau 1943, Reeve 1948, Fahn 1982). King and Heyes (1986) found difficulty in identifying the site of the stelar promeristem earlier than the fourth day of *Pisum sativum* embryo development. A similar situation was found in *Phlox* and certain other species (H. A. Miller and Wetmore 1945). Precursor cells of *Nicotiana* procambium appear at the 20-cell stage (Souèges 1920), and similar cells of *Trifolium minus* at the 120-cell stage (Johansen 1950). On the other hand, the procambium is only apparent shortly before the appearance of the cotyledons in a species of *Capsella* (Foster and Gifford 1974). Although this is a vitally important stage for the determination of future vascular elements, only a cytological description is available for the origin of procambial precursor cells.

1.3 General Mechanisms of Programming

Many questions concerning the initiation of vascularization may be raised. For example, what biochemical changes are occurring in these cells? Should we be looking for morphogenetic fields (Wardlaw 1952), and, if so, is the concept of pattern and information transfer relevant (Wolpert 1971, 1981)?

Several authors have stressed the importance of positional information (Holder 1979, Barlow 1984, J. M. Chapman and Perry 1987). The Wolpert (1970) model is simple and supports the general hypothesis that cell development is primarily determined by the morphological position the cell occupies in the organism as a whole. This model of positional information includes the following points: (1) In a developing system, cells have their positions specified in relation to one or more points within the tissue system. (2) Polarity refers to the points which specify the position of a given cell. Polarity is the direction in which positional information is specified or measured. (3) Positional information largely determines, with respect to cell genome and developmental history, the nature of the molecular differentiation that the cell will experience. The specification of positional information, in general, precedes molecular differentiation. (4) The same mechanisms that specify positional information can operate in different fields within the same system, and in completely different systems from different genera or even phyla. (5) Classical cases of pattern regulation show size invariance, which is largely dependent upon the ability of the cells to change their positional information in an appropriate manner and to interpret this change (Wolpert 1970).

At a very early stage of development the embryo contains information in the procambium specifying cytodifferentiation of vascular elements with specialized secondary cell walls (Gahan and Rana unpublished). Procambial cells of the early stage of embryonic development are determined (Meins and Binns 1979), that is, the program remains irrespective of any later programming events that occur at this time or subsequently. This rigidity of programming is associated with the procambium, whereas the adjacent procortex (periblem histogen) remains developmentally plastic as seen in adult plant roots (Rana and Gahan 1982, 1983 a). Therefore one can look for either a specific programming of a few cells, or for a more general field effect from which the programming of the procambium will remain after an adjustment to the initial programming in adjacent tissues by the formation of localized subfields. In assessing which of these possibilities might be employed, there are several lines of evidence that may be helpful. The presence of a biochemical marker permits the cytochemical identification of cells determined as stelar cells within the primary meristems of root and shoot apices of *Vicia faba* and *Pisum sativum* (Gahan 1981, Rana and Gahan 1982, Gahan and Bellani 1984). The enzyme marker is carboxylesterase (EC 3.1.1.1). This enzyme is resistant to the inhibitors eserine and p-chloromercuriphenylsulphonic acid. The enzyme, however, is inhibited in the presence of diisofluorochlorophosphate (Gahan 1981, Gahan and Bellani 1984). The carboxylesterase was associated with cell wall metabolism (Gahan 1981), and it may act by cleaving ester linkages between phenols and polysaccharides (Chandra et al. 1986). Application of a suitable cytochemical reagent for the identification of this enzyme in developing embryos of *Pisum sativum* has provided the following evidence: (1) The smallest embryo in the earliest stage of pod formation has a substantial number of cells giving a positive cytochemical test for the carboxylesterase marker and in excess of the region previously defined by Reeve (1948) as the stelar promeristem. (2) In earlier stages, at days 4–5 after anthesis (Lambert and Linck 1958), the embryo reacts positively for the marker enzyme in regions which are greater than those described by King and Heyes (1986). (3) Explants were prepared from embryos in all stages of development, from the smallest to the completely mature embryo in the pod. The small embryos were bisected and transverse sections were prepared from the larger ones, and the embryo fragments were cultured on a basal medium (Murashige and Skoog 1962) containing sucrose. These explants produced tracheary elements from the existing procambial initials or procambium, and all of the explants were apparently preprogrammed for vascular differentiation. In addition, carboxylesterase activity has been observed in early embryos of *Brassica* species and *Trifolium repens* and in mature embryos of *Lactuca sativa* and *Pinus sylvestris* (Gahan unpublished). Germination of the mature embryo of *Pisum sativum* reveals the presence of carboxylesterase activity in the procambium and cortex after 24 h of imbibition. Enzyme activity in the cortex was lost after 48–72 h of dark germination (Carmignac and Gahan unpublished). There appear to be two possibilities. The carboxylesterase activity of the cortex may be regulated genetically during the early hours of germination, thus separating the cortex from the other nonreactive tissues of the embryo. On the other hand, this enzyme regulation may have occurred prior to germination. One may be observing the loss of activity of existing carboxylesterase molecules. This enzyme is known to have a biological life span of at least 7 days (Dawson and Gahan 1979).

It has not yet been possible to identify either the number or function of the genes involved in the programming of such determined cells. The meager information available indicates the complexity of precisely identifying the number of genes involved, the expression of which is embryo-specific, let alone stelar promeristem-procambium specific genes (Dure 1985).

A well-defined procambium extends from the main axis of the embryo into the cotyledons of many spermatophyte seeds (Fig. 1.3). The cotyledon procambium may be derived subsequently from that of the main axis prior to germination.

A study of somatic embryos of *Daucus carota* var. S. Valery, using carboxylesterase activity as a marker for cell determination, gave clear evidence with regard to the time of initiation of procambial cells (Caligo et al. 1986). The somatic embryos of carrot were derived from cultures initiated from hypocotyl explants (Sung et al. 1984). The developmental stages (globular, heart, and torpedo) were examined by light microscopy using conventional histological methods. The early globular stage, isolated on the third day of culture, showed no carboxylesterase activity and no obvious cytodifferentiation. By the sixth day of culture the late globular stage showed carboxylesterase activity prior to any visible changes in the reacting cells. The enzyme activity indicated the sites for the subsequent development of cotyledons. The early heart-shaped stage showed activity associated with the cotyledonary traces, and enzyme activity was present in the hypocotyl in the late heart-shaped stage. The same reaction pattern was found in the torpedo stage (day 12), and esterase activity extended along the embryonic axis from the hy-

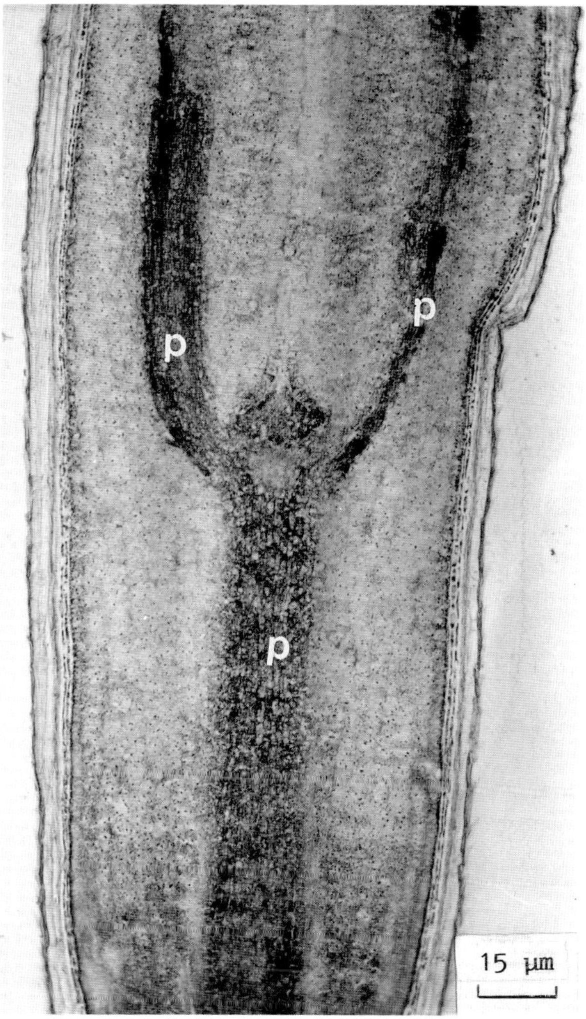

Fig. 1.3. Longitudinal, unfixed, frozen section of embryo of *Lactuca sativa* reacted for glucose 6-phosphate dehydrogenease activity (Gahan 1984) and showing the strongly reactive procambium (*P*) extending from the main axis and into the cotyledons

pocotyl to the root tip. Presumably polarity was established initially in the late globular stage, although provascular tissue was not visible at this time. By using serine and proline to block embryo development, the early globular stage has been shown to contain cells capable of differentiating into xylem. The xylem tissue was always confined to the middle of the globular structures, which failed to develop polarity (Nuti Ronchi et al. 1984, Caligo et al. 1986). The investigators concluded that there is a developmental stage during which determination of the morphogenetic pattern of the embryo is established, and that it is not impeded by the presence of serine and proline in the medium. In regard to vascular development, there are similarities between events observed in somatic embryos in vitro and in embryos in vivo. Additional research is required on both of these systems in order to identify the initiating factors and control of determination of the procambium.

1.4 Quiescent Center

The quiescent center (QC) is absent from the root apex of several species, for example, *Sinapis alba, Vicia faba, Malva sylvestris* at germination, but it forms before emergence of the radicle in *Pistia* and *Eichornia* and is formed once before and once after emergence in *Zea mays* (Clowes 1958, 1961, 1978, Bryne 1973, McLeod and McLauchlin 1974). When the QC is formed after germination, the radicle and plumule apical meristems are not always clearly defined (Fig. 1.4), and yet such a radicle primary meristem will already contain a central core of cells determined to form vascular tissue. The QC will be organized between this region and the root cap. The derivation of the QC is of interest, in roots of *Pisum sativum* and *Vicia faba*, because the carboxylesterase activity marking determined cells for vascular tissue differentiation is absent from the QC (Gahan and Rana 1985).

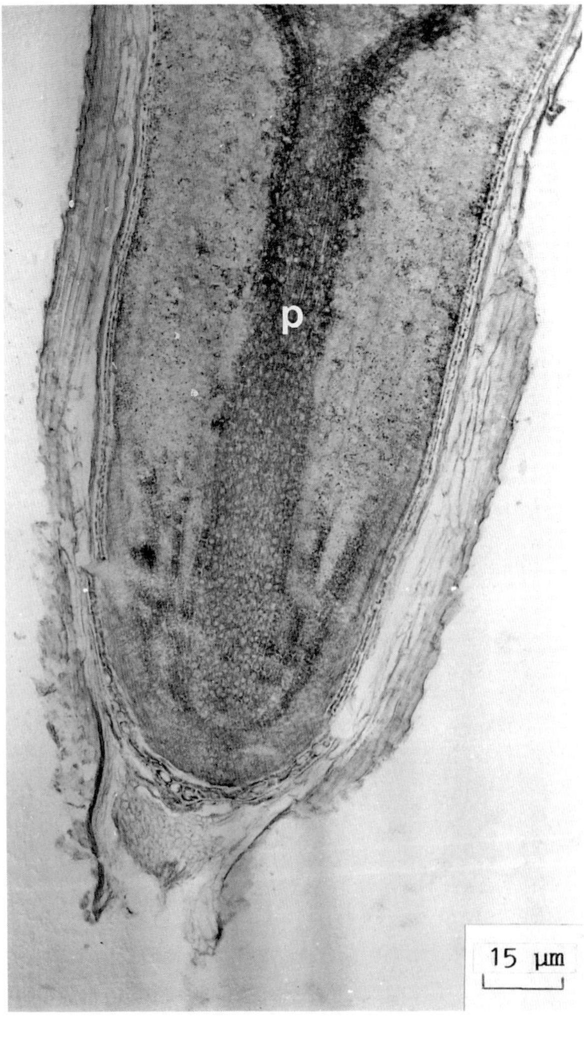

Fig. 1.4. Longitudinal, unfixed, frozen section of embryos of *Lactuca sativa* reaction for glucose-6-phosphate dehydrogenase activity (Gahan 1984), and showing the lack of an evident quiescent center *P* procambium

Hence either the QC is formed from the determined meristematic cells, and is reprogrammed along the lines of the procortex (see above), or it is derived from adjacent nondetermined cells. In view of the irreversibility of programming for stelar tissues (Gahan 1981, Rana and Gahan 1982), the latter explanation appears more likely. In this case, the cells produced by the QC in the root apex will be determined subsequently as either prostelar meristematic cells or as precursor cells of the root cap, cortex, and epidermis. Assuming the cells leave the QC determined as prostelar cells, do these cells need to synthesize carboxylesterase in order to differentiate further, or is there an activation of presynthesized enzyme molecules? If the prostelar cells leave the QC in an undetermined state, then they must be programmed subsequently. The latter programming may be influenced or regulated by previously determined cells adjacent to them (Barlow 1976, 1984). Assuming the enzyme is present in the QC, it is evidently in an inactive form. There is no response to cytochemical reagents for carboxylesterase in the QC, and attempts to label the enzyme with ^{3}H-diisofluorochlorophosphate have been equivocal (Gahan and Rana 1985).

1.5 Quantal Mitosis and Differentiation

In framing a hypothesis for the gene control of cytodifferentiation, it can be argued that gene activation requires that the progenitor cell be in a particular phase of the cell cycle, a topic which has been well reviewed (Roberts 1976, Shininger 1979 b, Phillips 1980, Dodds 1981 b, Fukuda and Komamine 1982 b). If cells leaving the QC are not programmed for vascular tissue differentiation, then they must become programmed during subsequent development. In some systems DNA synthesis and cell division appear to be prerequisites for the initiation of cytodifferentiation. Fosket (1968) was one of the first to suggest this requirement in connection with his experiments on induced xylogenesis in *Coleus*. Therefore, the question arises as to whether or not a quantal mitosis is needed at this stage of QC cell differentiation, or, indeed, in any plant cell that is scheduled for determination. The term quantal mitosis was proposed for a mitotic division that received signals for the initiation of a differentiation program (Holtzer 1970). Although the term was proposed for animal cells, Dodds (1981 a) suggested that the model of Vonderhaar and Topper (1974) for quantal mitosis might be applied to plant cells. According to this model there is an open "window" in G_1 when the cell can receive instructions for differentiation into a particular cell type (Fig. 1.5). Cells arrested early in G_1 will, on reentering the cell cycle, pass the window and receive information for cytodifferentiation without any further cell cycle activity. Or, if a cell is arrested late in G_1, that is, after the window, then it must complete a quantal mitotic division in order to return to the G_1 window. This is a difficult hypothesis to test. If cells pass through mitosis prior to differentiating, for example, parenchyma cells in explants from tubers of *Helianthus tuberosus* (Malawer and Phillips 1979) or cortical explants from pea roots (Shininger 1975 a, Simpson and Torrey 1977), it can be argued that one of three rounds of mitosis in *Helianthus tuberosus* or one of two rounds of mitosis in the case of peas, preceding tracheary element differentiation are quantal mitoses. Indeed, Shininger (1979 b) has ex-

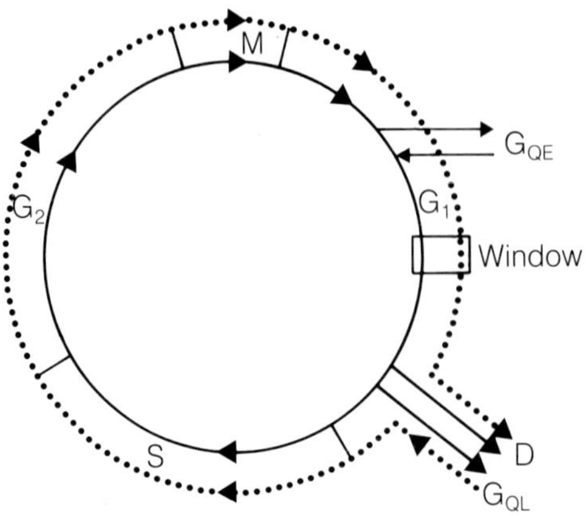

Fig. 1.5. Cell cycle activity and cytodifferentiation. Cells moving into and out of quiescence early in G_1 (G_{QE}) pass the window and differentiate (*D*) directly (*inner line*). Cells moving into and out of quiescence late in G_1 (G_{QL}) must complete the cell cycle in order to pass the window and differentiate (*outer line*). The latter sequence of events represents a quantal mitosis

pressed the view that DNA synthesis is a critical event in xylem differentiation. However, if a cell differentiates directly without mitosis, as in cultured mesophyll cells of *Zinnia* (Kohlenbach and Schmidt 1975), it can be argued that the cells were arrested early in G_1, passed the window, and received stimuli for differentiation. There are other examples of cells differentiating in the absence of a quantal mitosis. Thus, cytodifferentiation occurred in cultures of *Helianthus* following suppression of mitosis by gamma irradiation (Phillips 1981 b). Furthermore, inhibition of mitosis with colchicine was ineffective in blocking the differentiation of cortical parenchyma cells into a vascular bridge in wounded roots of *Pisum sativum* (Hammersley and McCully 1980). Equally, serial observations of isolated single mesophyll cells from *Zinnia* showed that about 60% of the total cell population formed tracheary elements without dividing, a process which was not stopped in the presence of colchicine (Fukuda and Komamine 1982 b). In addition, another group of cells underwent differentiation, but only after DNA synthesis, that is, in G_2.

In many systems it appears likely that cell division is necessary in order to increase tissue mass at sites where differentiation is occurring (Gahan and Rana 1985). It may be argued that there is no need for a quantal mitosis prior to cytodifferentiation in adult plant tissues, since there is no site which will be totally embryonic and beyond the influence of endogenous plant hormones. If such a site exists, it may be at the earliest stages of embryogenesis during the initiation of the first procambial precursor cells. In addition, evidence against the need for a quantal mitosis, involving programming in G_1, is that there is no G_1 phase in the cell cycles of some plant tissues. This occurs in *Zea mays* at the time of active root growth (Clowes 1971, Gahan and Hurst 1976). In fact, G_1 may not be an integral part of the cell cycle in eukaryotes and prokaryotes (Cooper 1979, 1981). The G_1 phase can be of variable length from zero to more than 100 h, and it represents a time interval for the preparation for DNA synthesis (Fig. 1.6). Therefore a G_1 phase exists only because the doubling time of cells is greater than the combined periods of $S+G_2$ (Fig. 1.6). It is possible that the G_1 phase is an expression of cell aging and therefore appears only after a certain passage of time (Hayflick 1965, 1977, Gahan and Hurst 1976, Mozaffari and Gahan 1978).

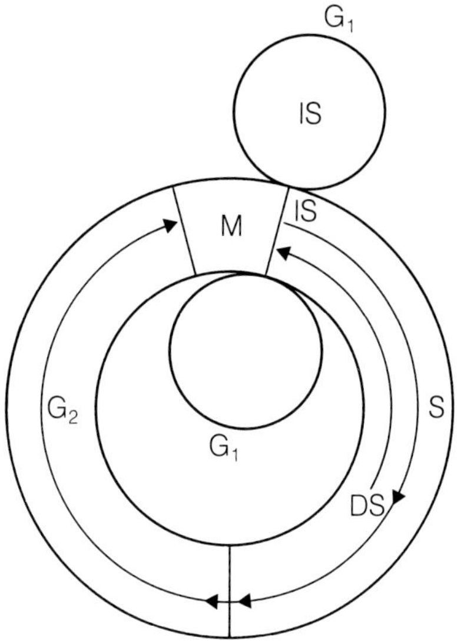

Fig. 1.6. Cell replication sequence when DNA synthesis (*DS*) is preceded by the synthesis of an initiator (*IS*). G_1 phase exists only when the cell doubling time is greater than $S+G_2$. *M* period of mitosis; *S* period of DNA synthesis

A strong argument against the concept of Cooper (1979, 1981), that G_1 may be an unnecessary part of the cycle, comes from studies of *Saccharomyces cerevisiae*. In this microorganism there is need of a gene in G_1 for commitment to the cell cycle and another gene in G_2 for the initiation of mitosis (Sudbery et al. 1980, Nurse and Bissett 1981). A gene (cdc 2) for the cell cycle start and control of mitosis has been isolated from *Schizosaccharomyces pombe*. Also a functionally homologous sequence which complements a cdc 2 mutation has been obtained from the gene bank of *S. cerevisiae*, and this sequence appears to contain the cdc 28 cell cycle start gene of *S. cerevisiae*. This indicates that the cdc 2 and cdc 28 genes perform homologous cell cycle functions in the two organisms (Beach et al. 1982). The implication is that such genes operate in other eukaryotes, for example, the *min* gene in barley (Bennett 1981). It is unclear how this might function in a cycle lacking a G_1 phase, as in the early stages of germination and growth of *Zea mays* (Gahan and Hurst 1976). Perhaps the method of cell cycle analysis by [^{3}H]-thymidine autoradiography is not sufficiently sensitive to detect a very short (minutes) G_1 period. Several classical cytogenetic studies have indicated that the initiation of chromosome replication can occur as early as the preceding telophase. This has been demonstrated by the telophase labeling by [^{3}H]-thymidine of the chromatin (La Cour and Pelc 1958), a mechanism by which the G_1 phase could become redundant.

1.6 Multiple Gene Copies and Xylogenesis

Cells in root apices from several species which are differentiating to form primary xylem elements have high levels of ploidy. Sets of chromosomes vary from 2c to 64c and even higher (Swift 1950, Deeley et al. 1957, Jensen 1958, McLeish and Sunderland 1961, Bucknall and Sutcliffe 1965, Corsi and Avanzi 1970, Bennett and J.B. Smith 1976, Lai and Srivastava 1976). This multiplication of chromosomes and genes appears to occur by endoreduplication (Tschermak-Woess 1956, Barlow 1978, Nägl 1978), although in *Marsilea quadrifolia* the cells undergo synchronous mitosis and become multinucleate (List 1963). The latter phenomenon also occurs in *Marsilea diffusa* (Vallade and Alabouvette 1985). In some cases all copies of particular genes appear to be expressed (Innocenti and Avanzi 1971, Avanzi et al. 1973, Nägl 1973, Durante et al. 1977). Not all plants show gene multiplication, and, in these cases, primary xylem elements are formed from diploid cells in root apices, for example, *Bryonia verrucosa* (Brabec 1953) and *Dennstaedtia circutaria* (List 1963). Diploid cells also produce tracheary elements under in vitro conditions in *Helianthus tuberosus* (Phillips 1981a). In addition, there is no confirmed polyploidy during primary xylem development in shoot apices. Since diploid cells are capable of producing xylem elements, it does not appear that polyploidy is a necessity for the induction of xylem differentiation.

What then is the role of endoreduplication in vascular tissue formation? Since a heavy demand is made to synthesize cell wall components and cytosol constituents prior to the loss of the nucleus, the additional gene copies could be important for the biosynthesis of long-lived mRNA and rRNA. This could provide the cell with the necessary information for synthetic processes in the subsequent absence of the nucleus. Since the lack of additional gene copies does not appear to be a limiting factor in diploid cells, perhaps polyploidy is useful but not mandatory. The number of gene copies that are actually used is difficult to determine, when one considers genes other than those found in the nucleolar organizer region of the chromosomes. The identification of active and inactive genes may be related to their position in the nucleus of the differentiating cell (Comings 1968, Heslop-Harrison and Bennett 1984). Analyses have shown that, by reconstruction of a nucleus or metaphase figure from serial sections, it is possible to map the location of each chromosome in relation to the position of the centromeres. The data reveal that (1) all centromeres have definite positions on the metaphase plate; (2) complete haploid genomes tend to be separated in diploid and hybrid cells from interspecific and hybrid species of grasses; and (3) heterologous chromosomes are in a predictable mean fixed order within each of the haploid genomes. The chromosome arms on heterologs which are adjacent in the order are closely matched in size (Bennett 1981, 1982, 1983). These data are similar to the positioning of the blocks of heterochromatin found along the arms of chromosomes from *Scilla* (Greilhuber and Loidl 1983). In terms of the differentiation process, if the chromosomes are in a fixed position, this implies that not only are all the genes on one chromosome linked, but they tend to be associated with the genes on the adjacent chromosome in the sequence. Assuming all the chromosomes are strictly ordered, then it is implicit that the genes will be similarly ordered. Bachman (1982) has shown that *Escherichia coli*, the DNA of which has been mapped in detail, contains clusters of the functionally related genes, for example, genes for chemotaxis, motility, and flagellar synthesis are all along the genome from the gene *thr A*. If this format holds for the chromatin distribution in the nuclei of higher plants, it may be helpful in trying to identify the genes associated with vascular tissue differentiation. Although there is some conflict in the literature (Heslop-Harrison and Bennett 1984), most workers favor the interpretation that genes being expressed are located in the outer 50% volume of the nucleus with the exception of the nucleolar organizer, which is within the internal 50% volume of the nucleus. This was seen especially in hybrid cells (Heslop-Harrison and Bennett 1984), when the nucleolar organizer regions of the inner chromosome complement were usually expressed, while those located in the outer chromosome complement were not expressed. Although this distribution of chromatin versus gene activity has been shown clearly for barley and maize, it has, nevertheless, important implications for studies on the expression of multiple gene copies in polyploid nuclei during xylem differentiation. The *min* gene, responsible for polyploidization in root cells of barley, occupies the end position in the outer chromosome chain. This locus may permit it to be at a controlling site for the determination of cell division (Bennett 1981).

Extra gene copies may be created whenever necessary without endoploidy. Several studies on root and shoot systems have shown the presence of a recently synthesized and unstable DNA fraction in cells differentiating to form vascular tissue (Pelc and La Cour 1959, Sampson and Davies 1966, van Parijs and Vandendriessche 1966, Brunori 1971, Hurst and Gahan 1975, Gahan 1976). This fraction has been termed "metabolic DNA" (Pelc 1972) to indicate amplification of extra gene copies for mRNA production, thereby permitting the master copy to remain in pristine condition. The additional slave copies are short-lived and this explains the reason for excessive replication of DNA during interphase in cells which normally do not divide. Although some situations have been identified as DNA repair sites, this is not the explanation for extra DNA in all cases, for example, in mammalian embryos (Dini and Viola-Magni 1976) and in seminal vesicles (Pelc and Gahan 1959). In the case of *Lycopersicon esculentum* the labeled DNA fraction of nondividing differentiating nuclei remains stable for several weeks before being lost (Hurst et al. 1973, Hurst and Gahan 1975, Gahan 1976).

1.7 Procambium to Cambium

Most of the current literature on the origin of the cambium concerns observations taken from shoots and petioles of a relatively few species, and there are few reports involving roots. Catesson (1984) has compiled a list of differences between the procambium and cambium based on Sterling (1946), Roland (1978), Larson (1982), and her own observations. The procambium possesses a single cell type which can be described as small and isodiametric in section. Cell division occurs in any direction, there is no intrusive growth, and the cell walls are of uniform thickness. Procambial cells contain proplastids, many small vacuoles, and are basophilic, staining intensely with pyronin for RNA. The cambium is composed of two cell types. Both long axial and short radial cells are present, and intrusive growth may occur. The radial walls are thicker than the tangential walls, and the chemical composition of each cell type is different. Each cell type, which stains weakly with pyronin, contains a large central vacuole; proplastids or chloroplasts are present.

Fahn et al. (1972) offered evidence, based on studies of wounded stems of *Ricinus communis*, that procambium and cambium were different developmental stages of the same tissue. The literature contains few specific descriptions of the phase change from procambium to cambium. This lack of information is partly due to the varied rates of development of the secondary tissues adjacent to the cambium, and, in addition, the system is complicated by seasonal fluctuations in activity (Larson 1982). By serial sectioning seedling shoot apices of *Populus deltoides*, Larson (1976) demonstrated that progressive development occurs from procambial strand → procambial trace → initiation layer → metacambium → cambium. Proto- and meta-xylem and phloem were formed from the procambial strand and metacambium, respectively, whereas secondary xylem and phloem were produced from the cambium. This information is of limited value in the current discussion, because the analysis of cambial activity and the hormonal regulation of this highly plastic tissue is restricted by a rigid description (Bannan 1950, Woch 1981, Catesson 1984). The reason for the emphasis on the anatomical description and identification of the cambial initials and the cambial zone is in relation to the functions and hormonal control of this meristematic tissue. In the case of long-lived species, cambial initials will divide hundreds or even thousands of times (Sachs 1981), and, in doing so, risk harmful mutations from mutagenic agents and from errors occurring during the mitotic cycles. Mitotic plant cells may be compared to dividing animal cells. Many animal cells appear to have a finite number of popuölation doublings that they are capable of achieving; for example, human skin and lung fibroblasts are limited to 50 ± 10 population doublings (Hayflick 1965). A finite number of cell divisions occurs in the root apical cell of *Azolla*, which divides 56 times during the production of cells required for the completion of the root structure. Similarly, there is a loss of mitotic activity in the apical meristems of *Vicia faba, Pisum sativum, Allium capa*, and *Zea mays* main roots, and during the cessation of division there is an increase in the number of aberrant mitoses (Mozaffari and Gahan 1978).

Catesson (1984) suggested that since a secondary meristem functions only in two planes, it is a mathematical certainty that there will exist a particular radial file of cells, at a given moment and place of origin, that are capable of producing derivatives first toward the phloem and subsequently toward the xylem. This radial file of cells consists of cambial initials. Such a definition is favorable to the discussion of the control of xylem and phloem differentiation by endogenous hormone levels and gradients which would operate over relatively short distances.

In terms of the capability of the cambium to divide many times (Sachs 1981), there is little specific information. It is of interest to examine an analogous condition existing in the small intestine of animals. The cells of the epithelium lining the villi are formed in the crypt region (Fig. 1.7). The nuclear DNA of cells in the crypt region were labeled following injection of tritiated thymidine intraperitoneally 1 h prior to the death of the animal (Leblond and Messier 1958). Mice killed at longer time intervals after injection of the label showed that the cells with labeled nuclei moved from the crypt to the tips of the villi. Although the epithelial cell initials reside at the base of the crypt, producing all the epithelial cells, all of the labeled cells migrated and after 12 h none could be found in the crypt. In other words, the whole labeled mitotic population of cells departed from the crypt, leaving behind a newly formed and unlabeled mitotic population of cells that continues generating new epithelial cells. This may be explained either by conservative

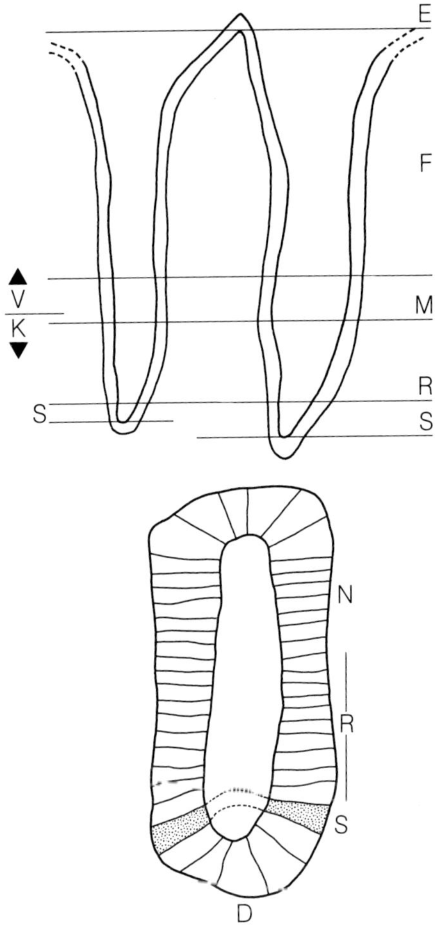

Fig. 1.7. Diagram of a transverse section of the small intestine of a mouse. **A** Villi lining the interior surface of the small intestine. Cell extrusion zone (*E*) functional pool of cells (*F*) maturation pool (*M*) proliferation pool (*R*) stem cells (*S*) villus (*V*) crypt (*K*). **B** Detail of crypt region. *N* non-dividing cells; *R* proliferating pool; *S* stem cells; *D* very slowly dividing cells. Cell cycle times are 24 h for S and 11 h for R

replication of DNA with the cell containing the newly synthesized DNA being the migratory cell, or by a dilution of the labeling through repeated mitoses in the absence of new sources of [^{3}H]-thymidine. A labeled nucleus with 100 silver grains in an autoradiograph needs only to divide an additional six times (the DNA being replicated semi-conservatively) for all labeling to be lost.

If there is difficulty in identifying cambial initials, possibly it is because these cells behave in a manner similar to the epithelial initials of the small intestine. In this case it may prove difficult to identify them in the cambia of plants other than annuals. The following example may offer an explanation as to how they avoid a high frequency of mitoses. Hurst and Gahan (1975) conducted an autoradiographic study of cut shoots of *Lycopersicon esculentum*. Cut shoots, five nodes in height, were fed with [^{3}H]-thymidine for 24 h. The label was shown to be incorporated into nuclear DNA. The second internode from the base of the shoot was analyzed, and 65.4% of the cambial cell nuclei were labeled after a 24-h exposure to the isotope. The cut shoots were then rooted and permitted to grow. The percentage of labeled cells rose to 86.9% 2 weeks after root initiation. There was a decline to 11% after 6 weeks, and no labeling was detected after 14 weeks. It would appear, from the decreased grain count distribution for each sampling, that cell division accounted for the increased labeling index seen at 14 days in comparison to the pattern after 24 h. Differentiation of the labeled cells to form phloem and xylem could be offered as an explanation for the loss of all the labeled cells from the cambial zone within 2–14 weeks. *All* of the labeled cells "migrated" from the cambial zone and so, in a manner similar to the intestinal crypt cells, there appears to be a complete turnover of cambial initials. This may also be indicative of the dilution of the labeling by repeated mitoses in the absence of new [^{3}H]-precursor. In part, this is seen also in the water fern *Marsilea vestita* during embryogenesis. There is an unequal partitioning of old and new DNA strands after fertilization between [^{3}H]-labeled male gamete DNA and unlabeled female gamete DNA. In particular, the male chromatin old (labeled) DNA passes primarily to those apical cells which will become "stem" cells (Kuligowski and Tourte 1979, Tourte et al. 1980).

Auxin appears to stimulate cambial cell replication and xylogenesis, whereas gibberellins have been implicated in phloem fiber formation (Stant 1963, Wareing et al. 1964, Digby and Wareing 1966, Aloni 1976b, 1978, 1979, Savidge and Wareing 1981b, Porandowski et al. 1982, T.J. Wodzicki et al. 1982). The mechanism by which this delicate control of cytodifferentiation occurs has posed some interesting problems of interpretation. One model suggested by Catesson (1984) was that the cambial derivatives are programmed during the division of the initials, that is, the external cells receiving competence to respond to either auxin or gibberellin. An alternative model was suggested by Savidge and Wareing (1981b). These workers considered that neither the centripetal nor the centrifugal cambial derivatives are determined with regard to their ultimate cytodifferentiation, but the endogenous hormones and other inductive stimuli influencing the fate of the cells are different on the xylem and phloem sides of the cambial region. Catesson (1984) suggested that a local modification could cause the initial cell to lose its role. The initiating role would then pass to one of the nondetermined derivatives. In this manner, there could be a rapid adjustment to produce either phloem or xylem. Such a model has implications for initials or shoot cell populations in which it is important that some cambial cells *must* retain their capability to subsequently act as initials. Obviously, the mechanism will vary according to seasons and is most tightly controlled during the active growth period. This would also ensure a rapid turnover of initials at the start of the new growth season, since the overwintering initials may have sustained damage and be eliminated at an early stage. This occurs in primary meristems during the resumption of embryo growth in *Pisum sativum, Zea mays*, and *Vicia faba* (Mozaffari and Gahan 1978). It is impossible that a similar barrier system that prevents primary procambium from differentiation operates in the cambium, although this is more evident in species possessing a distinct row of cells identified as initials (Bannan 1950, Newman 1956). In attempting to decide which of the models (Catesson 1984 vs. Savidge and Wareing 1981b) might operate, some support for the first model arises from the study of carboxylesterase activity in the procambium and cambium (Rana and Gahan 1983a). A similar carboxylesterase found in the procambium for vascular tissue determination is also found in the cambium. This finding supports the anatomical definition of a cambium as being a determined meristem (Fahn 1982). Therefore *all* cells of the cambial zone possess the genetic program to differentiate into vascular elements, and presumably the final switch is made in response to a local event. The switching substances are usually considered to be auxin, gibberellin, and cytokinin, although second messengers or intermediates of secondary metabolism may be involved. The substances initiating the programming for determination could arrive by either the xylem or phloem (Allen 1980) and diffuse laterally to the cambial zone. This model requires the cambium to be a sink in order that the triggering substances do not cross from one vascular tissue to another, since this could result in a random pattern of differentiation. There is an interesting report that coniferyl alcohol induces phloem cells to lignify in carrot tissues (von Wacek et al. 1954). The alternative model states that triggering substances arrive in the phloem, diffuse to the cambium, and then migrate into the xylem. It is equally possible that the cambium remains isolated from such activity; new cells induced to differentiate by localized activity arising from neighboring cells that are in some stage of differentiation. The differentiation of cambial cells may be analogous to the hypothesis of the initiation of floral induction, namely, that small electrical potentials may be associated with cells switching to produce either xylem or phloem elements (Fensom et al. 1965, Braton and Henry 1977, Morris 1980, Burgess and Linstead 1982, Goldsworthy and Rathore 1985, Rathore and Goldsworthy 1985, Montavon and Greppin 1986, Montavon et al. 1987). Plant hormones and second messengers may be initiating changes from a distance instead of acting directly on a given cell.

In considering the vascular cambium, a distinction should be made between programming for interfascicular development and activity of the intrafascicular cambium. The previous comments concern the latter phenomenon. In regard to the former, there are two opposing models for interfascicular cambial activity. Siebers (1971 a, b, 1972) advanced the view that the interfascicular cambium developed at an early stage during the development of the procambium and was independent of the intrafascicular tissue. On the other hand, other workers have suggested that the interfascicular cambium is dependent on homogenetic induction by the intrafascicular tissue (Fahn et al. 1972, Esau 1977). Rana and Gahan (1983 b) reasoned that if the Siebers model is correct, then the preprogrammed interfascicular tissue should give a positive reaction for carboxylesterase activity. The experimental results indicated that this model is incorrect. In *Trifolium repens, Ricinis communis, Arachis hypogea*, and *Pisum sativum* the carboxylesterase activity appeared first in the cells adjacent to the intrafascicular cambium. The enzyme activity appeared to spread across the interfascicular tissue from both sides of the vascular bundles.

1.8 Maintenance and Extension of the Vascular System

This topic was reviewed by Sachs (1981) in an analysis that was based on the interaction of internal programs that result in the patterned differentiation of vascular tissues. Since his analysis is primarily at the tissue level, it offers a restricted view of events at the cellular level of gene programming. Observations on the control of vascular form in the main axis and branches led Sachs to suggest that an inductive auxin flux follows and maintains the predetermined polarity of the tissues. With the creation of new polarities, the plane in which cells and vascular tissues develop is reoriented. The flux of signals, upon which differentiation depends, is derived from the various tissues of the plant axis (shoot to root). Of the various possible flux signals, the bulk of the evidence indicates that auxin is an important component of the signal. Given the complex contents of the phloem sap, the possible roles of other hormones should be considered. Gibberellin has been reported to play a role in phloem fiber formation (DeMaggio 1966, Aloni 1979, Savidge and Wareing 1981 b, Aloni 1987), and ethylene is known to synergize the induction of xylogenesis (Savidge and Wareing 1981 b, Roberts and A. R. Miller 1982, Aloni 1987 b). The need exists for screening a broad range of plant growth regulators, ions, sugars, and other molecules if we are to have a clearer understanding of the mechanisms by which the programming, maintenance, and expansion occurs.

The inductive flux follows and maintains a predetermined polarity of the tissue. Vessels and sieve tubes, induced by either young leaves or by experimental manipulation, follow the original polarity and develop toward the roots. There are in vivo situations in which normal polar induction does not occur, but a new polarity is established. One example is the production of a vascular cambium which results in a polarity at right angles to the main axis. Rays are formed from the cambium in response to radial signals, passing between differentiating vertical tissues. This process appears to rely on the ability of mature tissues to transport signals without the transporting tissue itself being influenced. This also infers the irreversibility of some vascular tissue differentiation.

The reason that vascular differentiation is limited to defined strands of cells has been considered by Sachs (1981). He proposed that "cells respond to the signal flux by gradually differentiating to become the preferred or facilitated channels from this flux along its original axis." This hypothesis is based on the possibility of a gradual autocatalytic process resulting in channeling the signals in a way so that strands of cells are differentiated to form vascular elements. Sachs argues for a given axis rather than a direction or polarity. In the case of angiosperm leaves, the changing patterns of growth centers result in vascular networks rather than polar "drainage systems" (Mitchison 1981). The main axis of angiosperms responds to polarity in the sense that the procambium initiated in the embryonic axis is set after the imposition of polarity on the embryo. The idea of strand maintenance may be reinforced, in the case of xylem, by the proposition that material from an autolysing cell is transferred to the contiguous cells in the strand. This material then acts to stimulate or initiate the next steps in the developmental sequence (Table 1.1, Sheldrake 1973, Sheldrake and Northcote 1968 a).

One of the most difficult problems involves the control of the formation of different tissues and cell types. There is the possibility that the procambial cells of the shoot apex are determined prior to the formation of the new leaf primordium with which the procambial cells are associated. If this is the case, then there must be factors other than the supply of auxin produced by the leaf primordium that are involved in the determination of vascular tissues (Gahan and Bellani 1984). Cytokinin may be a candidate for such a factor. Regardless of the initiating factors, it appears that auxin is operating on a strand of cells which already express a general program to form vascular tissue. This view does not question the importance of auxin in the differentiation process, but presses for a wider concept to include possible changes in advance of auxin action. Sachs (1981) has suggested that "signals for vascular differentiation are part of a general positive feedback between root and shoot tissues that controls many essential features of plant development." Not only might source-sink situations be implicated, but also long-distance signaling from shoot to root (auxin?) and from root to shoot (cytokinin?).

Chapter 2

Hormonal Aspects of Vascular Differentiation

L. W. Roberts

2.1 Some Characteristics of Plant Hormones

Early research on auxins led the American Society of Plant Physiologists to appoint a committee to propose some definitions. Plant regulators were defined as organic compounds, other than nutrients, which in small amounts promote, inhibit or otherwise modify any physiological process in plants (Tukey et al. 1954). The committee further recommended that plant hormones be defined as regulators produced by plants, which in low concentrations regulate physiological processes. Hormones usually move within the plant from a site of production to a site of action (Tukey et al. 1954). It was considered that growth regulation by auxin resulted from a physiologically effective auxin concentration acting on a responsive target tissue or organ (Thimann 1937, Leopold 1955). The early workers also viewed the regulation of physiological processes as an interaction between auxin and other growth-regulating compounds (van Overbeek 1959).

The five major groups of plant hormones include auxins, cytokinins, gibberellins, ethylene, and abscisic acid. Many, if not all, physiological processes appear to be governed by two or more of these classes of hormones either simultaneously or in some sequential order. In addition, there are other naturally occurring substances that appear to have regulatory properties. Although the following compounds have not achieved hormone status, they are capable, under certain circumstances, of altering the course of plant development: tricontanol (Laughlin et al. 1983), polyamines (Galston 1983), brassinosteroids (Cohen and Meudt 1983), batatasin (Hasegawa and Hashimoto 1975), pisumin (Hasegawa et al. 1983), *Eucalyptus* growth regulators (Paton et al. 1980), trigonelline (L. S. Evans and Tramontano 1981), and jasmonic acid (Yamane et al. 1981). Trewavas (1983) suggested that the nitrate ion has the characteristics of a plant hormone, but this view would be excluded on the basis of the original definition (Tukey et al. 1954). Hormones may initiate reaction sequences involving second messengers, for example, via the regulation of intracellular calcium ion levels (Hepler and Wayne 1985). One study suggested that adenosine 3′,5′-cyclic monophosphate is a second messenger for gibberellic acid and auxin (E. G. Brown and Newton 1981). Another possible second messenger is lectin. A positive correlation has been shown between the appearance of lectin and cellular differentiation (Barondes 1981, Kauss 1981, Meimeth et al. 1985).

A complication exists in the definition in regard to hormone transport from a site of biosynthesis to a target site (Leopold and Noodén 1984). Often a precursor or complex is the transported form, whereas a conversion to the active form occurs at or near the target site. For example, 1-aminocyclopropane-1-carboxylic acid accumulates in the roots of waterlogged tomato plants, is transported upward in the xylem, and is ultimately converted to ethylene in the leaves (Bradford and Yang 1980, Jackson 1985). Auxin is transported from the endosperm to the coleoptile tip in the form of a myo-inositol complex. The complex is hydrolyzed by the tip tissues, and the basipetal flow of IAA regulates cell elongation (Nowacki and Bandurski 1980). Although plant hormones are obviously transported, cell-to-cell movement should not be mandatory for a compound to be classified as a hormone (Roberts 1976). The physiological responses resulting from stress-produced ethylene are presumably experienced by the cell source and a few contiguous cell layers.

The assumption is made that each plant hormone binds to some receptor, and the hormone-receptor complex has a primary effect on some biochemical system at the target site. The evidence suggests that the receptors may be specific hormone-binding proteins (Venis 1985, Chadwick and Garrod 1986). Since a given hormone may have completely different effects during the various developmental stages of a given cell, the existence of multiple receptors has been postulated (Vanderhoef and Kosuge 1984). The presence of receptor sites reflects the sensitivity of a tissue to a given hormone, and this characteristic may be more important in regulating a process than the concentration of the hormone (Trewavas 1982, Weyers et al. 1987). Cross (1985) has examined the evidence for the elusive auxin receptor.

Our knowledge concerning the mechanism of action of any plant hormone is meager. In fact, a biochemical reaction that directly involves a hormone is not known (Street and Öpik 1984). Although the riddle of the primary mechanism of auxin-induced cell elongation remains unanswered, it is instructive to review some of the hypotheses. In the 1960s Key (1969) and others centered their investigations on the regulation of specific mRNAs and protein synthesis by auxin. Later, in the 1970s, the interest shifted to the acid growth theory (Rayle and Cleland 1970). Both of these concepts were incorporated into the proposal of Vanderhoef and Dute (1981), that is, auxin regulation may involve a rapid response arising from proton efflux followed by a second control on gene expression. Technical advances in the 1980s have resulted in the identification of selective mRNA changes after a brief exposure to auxin. Coleoptile sections of *Zea mays* were shown to contain higher levels of specific translatable mRNAs within 10 min after the exposure to auxin (Zurfluh and Guilfoyle 1982). Subsequent "late mRNA" changes (>2 h) are presumably also under auxin control. Theologis (1986) has proposed that the rapid mRNA induction is the primary auxin-regulated event within the cell. Gene transcription-translation products, including H^+-ATPase, are transported within secretory vesicles and ultimately proton efflux and cell elongation occur. The Theologis hypothesis explains the observation that auxin-induced acidification is inhibited by both cycloheximide and the fatty acid synthesis inhibitor cerulenin (Theologis et al. 1985). Agents that block the biosynthesis and exocytosis of auxin-induced products also prevent both auxin-induced proton efflux and cell elongation (Theologis 1986). Auxin control of gene expression has been reviewed by Guilfoyle (1986).

A problem that plagues investigators is the presence of endogenous hormones that are either present as contaminants in the tissue sample or are synthesized during the course of an experiment. Except in the case of ethylene, attempts to discover the presence of these growth factors are seldom made in experiments involving the application of exogenous hormones. Unidentified gibberellin-like compounds are present in lettuce pith explants used in xylogenesis experiments, and these endogenous compounds may have some significance in the process of cytodifferentiation (Pearce et al. 1987, Sect. 2.4).

Additional information on the hormonal control of vascular differentiation can be found in the reviews by Roberts (1976), Gresshoff (1978), Shininger (1979 b), Savidge and Wareing (1981 b), Sachs (1981), Savidge (1983 a, b), Lamb (1983), W.P. Jacobs (1984), Naylor (1984), Fukuda and Komamine (1985), Bolwell (1985), and Aloni (1987 b).

2.2 Auxins

Auxin is undoubtedly the major hormonal stimulant for both phloem and xylem differentiation (Thompson and W.P. Jacobs 1966, Aloni 1980, Sachs 1984, J. Warren Wilson and P.M. Warren Wilson 1984). Auxin is defined as a plant growth regulator which stimulates shoot cell elongation and resembles indole-3-acetic acid in its spectrum of physiological activity. The molecular basis for auxin action in any physiological process still remains an enigma. Isolated mesophyll cells of *Zinnia* first show signs of secondary wall thickening at 77 h of culture in an NAA-kinetin medium (Fukuda and Komamine 1980 a). It is not known precisely at what time auxin is acting during this 77-h period. Auxin may have regulatory functions in RNA and protein synthesis between 24 and 60 h of culture (Fukuda and Komamine 1983), and this possibly involves either the synthesis of enzymes or tubulin required during cytodifferentiation. Changes in peroxidases (Masuda et al. 1983) and lignin-related enzymes (Fukuda and Komamine 1982 a) are not detected until later during xylogenesis, that is, after 77 h. Because the inhibition of PAL activity and lignification does not prevent secondary wall formation during xylogenesis, lignification does not appear essential for the initiation of xylogenesis (Smart and Amrheim 1985, Chap. 5.4). When auxin-induced xylogenesis in *Zinnia* mesophyll cells is suppressed by 2,3,5-triiodobenzoic acid, an auxin transport inhibitor, the cells develop an unusual polar swelling (Burgess and Linstead 1984, Chap. 4.6). This swelling probably results from a localized accumulation of auxin in the vicinity of the inhibited auxin carrier. Localized influx of ions, stimulated by the trapped auxin, may initiate vesicle secretion associated with membrane and cell wall enlargement (Naylor 1984).

Depending on the tissue involved and other circumstances, auxin can stimulate cell division, cell elongation, and cell differentiation. Bandurski and Nonhebel (1984) have proposed that auxin combines with a single cell receptor, and the multiplicity of responses attributed to auxin arise from the roles of the second messengers. A unified theory of hormone action involving calmodulin as a second messenger has been proposed (Elliott et al. 1983, Table 2.1). Other workers have shown that calcium salts have a modifying effect on each of the five groups of plant hormones (Poovaiah and Leopold 1976 a, b, Leopold 1977). The importance of calcium in the regulation of plant development has been reviewed (Hepler and Wayne 1985).

What evidence links calcium with vascular differentiation? Histological changes in *Pisum* root tips, resulting from calcium deficiency, were described by

Table 2.1. Comparison of the IC_{50} values for trifluoperazine and tetracaine in a number of plant hormone- and red light-stimulated biological systems. (Elliott et al. 1983)

Biological system	TFP µM	TC µM
Red light-dependent betacyanin synthesis	70	300
Cytokinin-dependent soybean callus growth	70	500
Cytokinin-dependent betacyanin synthesis in *Amaranthus* half-seedlings	150	600
Cytokinin-dependent cotyledon growth	50	ND[a]
Cytokinin-dependent betacyanin synthesis in *Amaranthus* cotyledons	150	ND
Auxin-dependent coleoptile growth in the absence of sucrose	50	800
Auxin-dependent coleoptile growth in the presence of sucrose	25	600
GA_3-dependent α-amylase induction (Triton X-100 extracted)	400	1000

[a] ND, not determined.
IC_{50}, concentration of calmodulin-binding drug giving 50% inhibition.
TFP, trifluoperazine; TC, tetracaine.

Sorokin and Sommer (1929, 1940). The calcium-deficient cells exhibited abnormal cell division and were incapable of initiating cytodifferentiation. Another study on the effects of calcium deficiency in the apical meristem of *Pinus taeda* indicated that xylem and phloem differentiation were markedly inhibited (Davis 1949). Some of the possible roles for calcium in vascular element differentiation include microtubule physiology, auxin transport, cell wall formation, and lignification.

Microtubules are necessary for the orientation of cellulose microfibrils during secondary wall formation (Hardham and Gunning 1979). Calmodulin is required for the assembly-disassembly of microtubules (Marcum et al. 1978, Hepler and Wolniak 1984), and both calcium and adenosine 3′,5′-cyclic monophosphate appear to be involved (Dedman et al. 1979). Although tubulin polymerization in vitro requires a calcium-free medium, calcium has certain regulatory functions in cellular events involving microtubules (Sabnis and Hart 1982, Hepler and Wolniak 1984, Hepler and Wayne 1985). Dimethyl sulfoxide promoted the assembly of cortical microtubules (Himes et al. 1977, Hahne and Hoffmann 1984). Nondividing isolated protoplasts were stimulated to divide, as well as initiate xylogenesis, by the application of dimethyl sulfoxide (Hoffmann et al. 1984, Hahne and Hoffmann 1984). Protoplasts of *Nicotiana glutinosa*, treated with colchicine, lacked microtobules and were incapable of division. When dimethyl sulfoxide (2% v/v) was added to the cultures containing the colchicine-treated cells, microtubules were found and the cells resumed division. Increasing the concentration of dimethyl sulfoxide to 4% (v/v) induced xylogenesis (Fig. 2.1). In another study it was found that the induction of lateral root formation in segments of radish (*Raphanus sativus*) was greatly enhanced by a pulse treatment of dimethyl sulfoxide (5% v/v) during the early hours of culture (Blakely et al. 1986). Since dimethyl sulfoxide increases membrane permeability, this solvent may stimulate the onset of these physiological events by increasing auxin uptake or possibly by altering intracellular calcium levels.

Vascular differentiation is a polar phenomenon involving auxin transport from cell to cell (Sachs 1984, 1986). The basipetal secretion of IAA has a specific

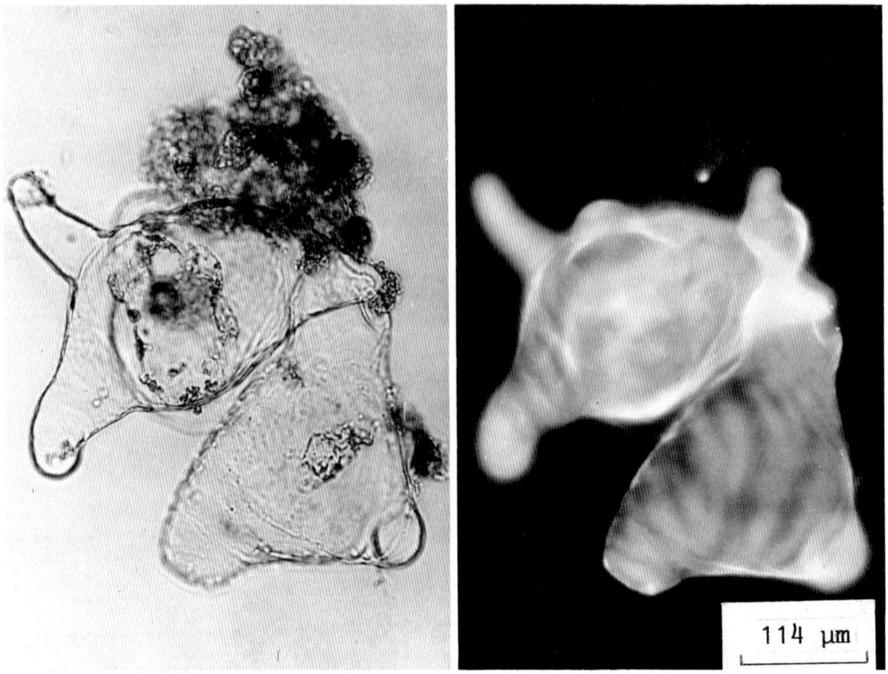

Fig. 2.1. Formation of "Viking helmet" structure by callus protoplasts (*Nicotiana glutinosa*) following culture in a medium containing colchicine and dimethyl sulfoxide (*left*). Fluorescence of the calcofluor white reagent reveals the banded pattern of cellulose microfibrils associated with xylogenesis (*right*). (Hoffmann et al. 1984)

calcium requirement (De Guzman and Dela Fuente 1984, Dela Fuente 1984). Calcium and auxin move simultaneously in opposite directions, and the movement of both substances is blocked by 2,3,5-triiodobenzoic acid (De Guzman and Dela Fuente 1984). This inhibitor also blocks xylogenesis (Roberts 1976, Burgess and Linstead 1984). According to Hertel's (1983) hypothesis, the exit carrier for the auxin anion is coupled with calcium antiport, that is, a counter flow of Ca^{2+} into the cell. Potassium ions may also be involved in charge balancing, since K^+ has been reported to accumulate in differentiating metaxylem elements (McCully et al. 1987).

Calcium is closely associated with cell wall metabolism. A preliminary report by Griffing and Ray (1979) indicated that calcium was necessary for the fusion of Golgi vesicles with the plasmalemma. The role of calcium in this process has been examined by several groups (Toister and Loyter 1973, Dahl et al. 1978, Baydoun and Northcote 1980a, b, 1981). Evidence of this requirement is illustrated by tip growth of pollen tubes. Disturbance of the normal calcium relations with the Ca^{2+} specific ionophore A23187 resulted in the irregular deposition of primary wall constituents and inhibition of growth (Reiss and Herth 1978, 1979). Calcium is required for the biosynthesis of both cellulose (β-1,4-glucan) and callose (β-1,3-glucan) (Delmer et al. 1984, Table 2.2). Since callose is synthesized

Table 2.2. Effectors of glucan synthetase activity

Effectors present	66% Ethanol-insoluble glucan	Alkali-insoluble glucan
	nmol per hour per mg protein	
None	7	4
CB	52	5
Mg^{2+}	5	22
Ca^{2+}	13	3
Mg^{2+}, CB	36[a]	22[a]
Ca^{2+}, CB	333	4
Mg^{2+}, Ca^{2+}, CB	400	213
Mg^{2+}, Ca^{2+}, GTP	13	17
Mg^{2+}, Ca^{2+}, GTP, CB	383	155

An enriched plasma membrane fraction isolated from cotton fibers was assayed for UDPGlu: glucan synthetase activity in the presence of celliobiose (CB), Ca^{2+}, Mg^{2+}, and guanosine-5′ triphosphate (GTP) in various combinations. Enzyme activity was dependent on a synergistic interaction between Ca^{2+} and CB for ethanol-insoluble glucan, whereas the alkali-insoluble glucan required a combination of Ca^{2+}, Mg^{2+}, and CB. The addition of GTP had no effect on glucan activity (Delmer et al. 1984).

[a] In some experiments these values were higher by a factor of two to three times those given.

during sieve plate formation, evidently the cytodifferentiation of phloem elements is dependent on intracellular calcium levels.

The possibility that calcium, associated with the cell wall, influences lignification by releasing wall-bound peroxidase has been examined by Wardrop (1976). A relationship between calcium, binding of peroxidase to the wall, and lignin formation was indicated by early workers (Lipetz 1962, Lipetz and Garro 1965). On reexamining the problem, Wardrop found that the Ca^{2+} level in unlignified cells was, in some cases, 20-fold the concentration present in the walls of adjacent cells. Calcium levels in the cell wall decreased during lignification. According to Wardrop (1976), these observations are inconsistent with the concept that calcium plays a direct role in the process of lignification. Westermark (1982) proposed an alternative idea concerning the role of calcium during lignification. The reaction of enzymatically produced superoxide radical with coniferyl alcohol was studied. Although no reaction occurred in the absence of calcium, coniferyl alcohol was oxidized by the superoxide radical in the presence of calcium. It was suggested that the superoxide-generating enzyme outside of the cell wall is responsible for lignification, and that the relatively high level of calcium in the adjacent nonlignified cells is required for the process to occur. Gaspar et al. (1985) proposed that the regulation of developmental processes involves the activities of basic and acidic peroxidases. Increased intracellular levels of calcium activates the secretion of basic peroxidases into the free space (Penel et al. 1984), and calcium promotes the binding of these basic peroxidases to membranes (Penel et al. 1979). A role for membrane-bound peroxidases acting as ACC oxidase in the conversion of ACC to ethylene has been suggested. In this hypothesis the main roles of ethylene would be in the regulation of PAL and acidic peroxidases associated with lignification. Acidic peroxidases are also secreted and activated by calcium according to Gaspar's group. The possible interrelationships between isoperoxidases, lignification, and cell differentiation have been discussed by Gaspar et al. (1982).

Assuming calcium is acting as a second messenger, the crucial question becomes: How does auxin regulate intracellular calcium levels? Intracellular free Ca^{2+} is very low (Williamson 1981). Hormonal or environmental stimulation of Ca^{2+} influx results in the activation of calmodulin. Since calmodulin stimulates the membrane-bound ATPase transporting Ca^{2+} into microsomes (Dieter and Marmé 1980), this is one way of regulating cytosol Ca^{2+} levels. Another control involves the efflux pumping of Ca^{2+} by plasmalemma-bound ATPase. This enzyme is greatly stimulated by auxin in membrane preparations taken from the elongation zone of soybean hypocotyl (Kubowicz et al. 1982). Also the transport of Ca^{2+} via the efflux pump serves to increase the electrical potential difference across the plasmalemma. Membrane potential serves a role in the synthesis of β-glucans (Bacic et al. 1980). Another interesting connection between calcium and auxin is the observation that auxin-induced proton secretion is dependent on the presence of exogenous calcium (Cohen and Nadler 1976), and, in fact, stimulates proton release (Hepler and Wayne 1985). Experimental evidence supports the hypothesis that auxin-induced cell elongation requires calmodulin. Calmodulin-binding compounds were highly effective in blocking auxin-dependent wheat (*Triticum aestivum*) coleoptile growth (Elliott et al. 1983) and auxin-induced elongation of oat (*Avena sativa*) coleoptile segments (Raghothama et al. 1985). Other workers have found that the induction of xylogenesis in cultured explants of Lactuca also requires calmodulin (Roberts and Baba 1987). The calmodulin-binding drugs trifluoperazine and chlorpromazine completely blocked auxin-induced xylogenesis, although callus formation was only slightly inhibited under the same cultural conditions. Chlorpromazine sulfoxide, an analog of chlorpromazine that is ineffective in blocking calmodulin, did not inhibit xylogenesis. Verapamil, an inhibitor of voltage-regulated Ca^{2+} transport, had little or no effect on xylogenesis. Also the incorporation of lanthanum ($LaCl_3$) had no statistically significant effect on xylogenesis. The movement of lanthanum is restricted to the apoplast (Thomson et al. 1973, Nagahashi et al. 1974), and La^{3+} is a competitive antagonist of Ca^{2+} at cell wall binding sites (R. B. Martin and Richardson 1979). These results support the view that apoplastic calcium is not a factor in auxin-induced in vitro xylogenesis (Roberts and Baba 1987).

In addition to calcium, another element closely associated with xylem differentiation is boron (Lewis 1980). According to Dugger (1983), the primary effects of boron is the regulation of physiological processes arising from complex formation between boron and hydroxy-rich compounds in which two hydroxyl groups are in the cis position. Complex formation presumably alters the rates of enzymatic reactions, produces shifts in metabolic pathways, and changes membrane characteristics. Boron is the only element known to be essential to higher plants that has no role in the animal kingdom (Parr and Loughman 1983). Aside from some species of the Bacillariophyceae (diatoms), the only plant taxon with a universal requirement for boron is the Tracheophyta. A deficiency of boron usually results in an accumulation of phenolic compounds and increased lignification. One hypothesis is that the borate ion regulates the partitioning of glucose between glycolysis and the pentose shunt pathway by complexing with 6-phosphogluconate (S. Lee and Aronoff 1967). The latter complex inhibits the functioning of 6-phosphogluconate dehydrogenase. This enzyme is in a pivotal position for the operation of the pentose shunt pathway, and, in turn, the shikimic acid pathway. A deficiency of boron would result in an abnormally high activity of this dehydrogenase and in the production of an excess of phenolic compounds. This scheme is complicated, since some dihydroxy phenolic acids produced by shikimic acid metabolism also form complexes with the available boron (Wardrop 1981). Objections have been raised against the acceptance of the Lee and Aronoff hypothesis (Parr and Loughman 1983). Another proposal is that boron is essential to lignin biosynthesis by regulating the hydroxylase and oxidase activities of phenolases involved in the formation of caffeic and hydroxyferulic acids (Lewis 1980). Some workers have suggested that boron deficiency leads to supraoptimal auxin levels in plants (Dugger 1983). Observations from Torrey's laboratory are in opposition to this view. The application of supraoptimal auxin to sunflower root tips did not elicit the same ultrastructural abnormalities that were produced by a deficiency of boron (Hirsch and Torrey 1980). Within 6 h after transfer to a boron-deficient nutrient medium, several physiological and ultrastructural changes had occurred. Using a radioimmunoassay technique for IAA, no significant difference was found in free IAA between boron-deficient and control root tips following 6, 24, and 72 h (Hirsch et al. 1982). The importance of boron in ion transport and membrane integrity has been considered (Parr and Loughman 1983). Other lines of evidence indicate that boron plays a role either in maintaining an adequate supply of one or more pyrimidine nucleotides or in the metabolic utilization of these compounds (Lovatt 1985). Lewis (1980) has pointed out some interesting possible relationships between boron, auxin, lignification, peroxidase activity, and ethylene. A comprehensive review of the various relationships of boron to plant metabolism is available (Dugger 1983).

Auxin induction of xylogenesis may be mediated via derepression of certain genes leading to RNA and protein synthesis. Early evidence supporting this mode of auxin activity was reviewed by Key (1969). Specific mRNA sequences arise following the treatment of pea and soybean segments with auxin (Theologis and Ray 1982, Zurfluh and Guilfoyle 1982). The nature of the enzymes encoded by these mRNA sequences are unknown (Vanderhoef and Kosuge 1984).

2.3 Cytokinins

Cytokinins are plant growth regulators that stimulate cell division and resemble 6-furfurylaminopurine (kinetin) in physiological activity. Cytokinins are mainly N^6-substituted aminopurines. Although it has been impossible to prove that a cytokinin-dependent event is necessary for the initiation of auxin-induced differentiation, the evidence is clearly in favor of some role for cytokinin in the formation of xylem and phloem elements. The earliest reports on the synergistic effect of kinetin and auxin on xylogenesis involved excised pea stem segments (Sorokin et al. 1962, Sorokin and Thimann 1964) and a tobacco tissue culture (Bergmann 1964). The presence of kinetin was found to block xylogenesis during the culture of *Coleus* stem slices (Fosket and Roberts 1964). This inhibitory effect was probably due to a supraoptimal concentration of cytokinin, since endogenous cytokinin is present in the xylem sap of this plant (Banko et al. 1976). Numerous in vitro systems have a dual requirement for auxin and cytokinin in the initiation of xylogenesis: suspension cultures of root callus of *Pisum* and *Convolvulus* (Torrey 1968), stem callus of *Centaurea* (Torrey 1968), soybean callus (Torrey 1968, Fosket and Torrey 1969), *Nicotiana* callus (Bergmann 1964), pea root segments (Torrey and Fosket 1970), pea root cortical explants (Phillips and Torrey 1973, Shininger and Torrey 1974), pith explants of *Lactuca* (Dalessandro and Roberts 1971), and *Zinnia* mesophyll cell cultures (Fukuda and Komamine 1980b).

There are some reports of tissue cultures initiating auxin-induced xylogenesis without the addition of cytokinin to the medium. Most of these systems probably contain endogenous cytokinins. Linstedt and Reinert (1975) found that carrot callus synthesizes cytokinin during culture on a 2,4-D-containing medium. The 2,4-D-induced biosynthesis of zeatin compounds in carrot cultivars is light-dependent (Mizuno and Komamine 1978, Chap. 5.4). Cytokinin biosynthesis may occur during 2,4-D-induced xylogenesis in lettuce pith explants (Dalessandro and Roberts 1971). Although exogenous auxin was the only hormone required for xylogenesis in explants of *Helianthus tuberosus* tuber (Dalessandro 1973), the tuber of this plant contains endogenous cytokinin (J.P. Nitsch and C. Nitsch 1960). Other examples of auxin-induced xylogenesis in the presumed absence of a cytokinin are more difficult to explain, for example, in pith cultures of tobacco (Clutter 1960) and *Coleus* (Earle 1968). Pith parenchyma tissue is not a cytokinin source in higher plants. After 32 days of culture Clutter (1960) found that both cell division and vascular nodule formation had occurred in her cultures. Since cell division in isolated tobacco pith explants requires the combined presence of auxin and cytokinin (C.O. Miller et al. 1956, Fosket 1977), the same hormone combination may be necessary in order to stimulate cellular proliferation in other isolated pith tissues. Based on this assumption, cytokinin biosynthesis probably occurred during the prolonged culture period (Fosket 1980). In the experiments conducted by Earle (1968) there was no indication that the pith explants were rinsed following the removal of the surrounding vascular tissues, and the isolated pith may have been contaminated with endogenous cytokinin released by the vascular tissues. Another unusual case involves the observation that ethylene substitutes for cytokinin in auxin-induced xylogenesis in pith explants of *Lactuca* (A.R. Miller et al. 1984, Chap. 5.5). Xylogenesis induced by IAA-kinetin in this system was different compared to xylogenesis occurring in the presence of either IAA-ACC or IAA-CEPA. Total numbers of xylem elements were always considerably greater with auxin-kinetin than with the ethylene precursors, and the location of the differentiated cells was different. In the auxin-kinetin treated cultures the newly formed xylem elements were mainly in the tissue adjacent to the culture medium, whereas in the ACC- and CEPA-treated material xylogenesis occurred near the upper surface of the callus distal to the medium. A third difference was the timing of the differentiation, that is, relatively few xylem elements differentiated until nearly 21 days of culture in the explants receiving ACC and CEPA (A.R. Miller et al. 1984). Because of the prolonged length of time required for the initiation of xylogenesis, cytokinin biosynthesis may have occurred.

Although the mode of action of cytokinins is unknown, there are several possible roles for this hormone in the reprogramming of cells for cytodifferentiation. Developmental botanists have been interested in cell cycle activity as it may be related to cytodifferentiation (Dodds 1981 a, b, Yeoman 1981), and the linkage might be via cytokinins. Although tracheary elements often differentiate after one or more rounds of cell cycle activity (Malawer and Phillips 1979), cell division is not a necessity for the initiation of xylogenesis in vitro. Numerous workers have observed the direct differentiation of *Zinnia* mesophyll cells (Chap. 4.6). Increasing concentrations of cytokinin apparently have a greater effect on stimulating xylogenesis than on increasing the rate of cell division in soybean callus (Fosket and Torrey 1969) and in cortical parenchyma of pea (Shininger and Torrey 1974).

Our best evidence is that cytokinins play one or more roles in protein synthesis. Cytokinins bring about qualitative as well as quantitative changes in protein synthesis, and, during the culture of soybean cells, these changes occur prior to cytokinin-induced cell division (Fosket and Tepfer 1978). Cytokinin appears to specifically regulate events either in G_2 or during the transition from G_2 to mitosis (Fosket 1977). In cultured cells deprived of cytokinin the cells either accumulate in G_2 briefly and then stop cycling, or the cells exhibit endoreduplication of DNA without mitosis or cytokinesis. Cytokinin may be necessary for the translation of certain genetic messages into proteins, which are functional during G_2, in order to complete the cell cycle (Fosket 1977). Translation could be promoted either by an activation of polyribosomes or by increasing their number via an enhancement of ribosomal binding into polysome configurations (Szweykowska et al. 1981). A recent study was conducted on the effects of two cytokinins [N^6-benzyladenine and N_1-(2-chloro-4-pyridyl)-N_2-phenylurea] on RNA synthesis in isolated nuclei removed from pumpkin (*Cucurbita pepo*) cotyledons (Ananiev et al. 1987). Within 4–6 h both cytokinins resulted in an increase of nuclear RNA-polymerase-I (EC 2.7.7.6) activity. This enzyme is responsible for rRNA biosynthesis. The increased activity was the consequence of an increase in the polyribonucleotide elongation rate. Cycloheximide inhibition indicated that de novo protein synthesis was necessary for the cytokinin-stimulated RNA polymerase activity (Ananiev et al. 1987).

Nucleic acid metabolism and protein synthesis have been examined in the *Zinnia* system by Komamine's group (Chap. 4.6). Replication of a whole genome during the S phase is not a prerequisite for xylogenesis (Fukuda and Komamine 1981 a). The peak of inhibition of xylogenesis by aphidicolin, a DNA synthesis inhibitor, occurred between 48 and 60 h of culture. This time period is after the S phase, but prior to any visible signs of wall thickening (Komamine and Fukuda 1982). Studies are in progress on gene amplification (differential DNA replication) during G_2 by the developing xylem elements of *Zinnia* (Komamine personal communication). In addition, it has been shown that the synthesis of RNA and protein occurring between 24 and 60 h of culture is correlated with xylogenesis (Fukuda and Komamine 1985), and cytokinins may be functional during these synthetic processes. The nature of the proteins synthesized remains unknown, although one may speculate that they are either isoperoxidases associated with lignification (Masuda et al. 1983) or enzymes involved in the biosynthesis of wall components (Northcote 1982, 1984, 1985 a).

In cultured soybean cells the synthesis of a 55000 Da protein was strongly promoted by cytokinin, and this approximates the molecular weight of the subunits of tubulin (Fosket et al. 1981). Tubulin is the major component of microtubules which are necessary for the formation of the mitotic spindle and phragmoplast (Hepler and Wolniak 1984). Microtubules also serve a regulatory role in the orientation of cellulose microfibrils in the developing secondary wall (Hepler and Palevitz 1974). Additional experiments have not supported the hypothesis that cytokinin regulates tubulin biosynthesis (Fosket et al. 1981).

As mentioned previously (Sect. 2.2), a hormonal theory has been proposed with intracellular calcium and calmodulin acting as second messengers. Evidence from several sources indicates that this theory may be valid for cytokinin as well as auxin (Hepler and Wayne 1985). Calmodulin-binding compounds blocked cytokinin-induced betacyanin biosynthesis in *Amaranthus* seedlings (Elliott et al. 1983, Table 2.1). Membrane-bound Ca^{2+} was localized at the presumptive bud site in the caulonema cells of *Funaria* 12 h after the application of cytokinin (Saunders and Hepler 1981). In fact, buds were induced by culturing the protonema in the presence of the ionophore A23187 plus Ca^{2+} in the absence of exogenous cytokinin (Saunders and Hepler 1982). These bud initials, however, were incapable of further development, and it appears that cytokinin provides some additional stimulus for bud development aside from the regulatory role of calcium. The treatment of meristematic or maturing, but not elongating, zones of soybean hypocotyl with cytokinin resulted in an increase in ATP-dependent Ca^{2+} transport in plasmalemma preparations (Kubowicz et al. 1982).

Are there other naturally occurring cytokinins or cytokinin-like compounds that might qualify as xylogenic hormones? Dihydroconiferyl alcohol (DCA) found in sycamore maple (*Acer pseudoplatanus*) sap and in maple syrup acts synergistically with kinetin to stimulate soybean callus, and the compound is also active in tobacco callus and radish leaf senescence cytokinin bioassays (T. S. Lee et al. 1981). Since DCA has been isolated from differentiating xylem of *Pinus contorta*, some experiments were performed to test its xylogenic properties (Savidge 1987). DCA was applied in lanolin to transverse apical surfaces of disbudded-defoliated cuttings of 1-year-old *Pinus contorta*. The branch cuttings had been excised in late winter and stored for 2 months at 4 °C with leaves and buds intact. The DCA treatment stimulated both cambial cell division as well as the differentiation of tracheids, whereas the lanolin-treated controls showed no signs of cell division or cytodifferentiation. The experiment was repeated with freshly collected disbudded-defoliated cuttings, and DCA had no visible effect on either cell division or xylem formation (Savidge 1987).

Dehydrodiconiferyl glucosides, isolated from *Vinca rosea* crown gall tumors, promoted cell division of tobacco cells (Lynn et al. 1987). The enantiomeric isomers A and B stimulated cell division, but they were required in concentrations up to two orders of magnitude higher than zeatin riboside in order to promote a comparable growth response (Binns et al. 1987). The research group engaged in these studies hypothesized that endogenous cytokinins may exert control of cell division by the production of these compounds. Since the dehydrodiconiferyl glucosides are epimeric phenyl propanoids, they are thought to be associated with cell wall metabolism (Binns et al. 1987).

2.4 Gibberellins

Unlike cytokinins, which stimulate mitotic activity in the presence of auxin in tissue cultures, gibberellins do not have a pronounced effect on increasing cell number of suspension cultures (R. L. Jones and MacMillan 1984). Exogenous gibberellic acid (Ga_3) did not stimulate cell division in cultures in which endogenous gibberellin biosynthesis had been blocked (Rappaport 1980). On the other hand, GA_3 is known to have a stimulatory effect on cell division in shoot meristems (Loy 1977). One reported effect is the initiation of the S phase in cells arrested in G_1 (Jacqmard 1968). In watermelon seedlings GA_3 greatly reduced the duration of the cell cycle by shortening the length of time the cells spent in G_1 (30% less) and in S (36% less) (P. B. W. Liu and Loy 1976).

During the 1960s, workers from several different laboratories reported that combinations of IAA and GA_3 had a synergistic effect on cambial activity and on xylem and phloem differentiation in woody dicots (Wareing et al. 1964, Roberts 1976). Opinions are divided, however, on the relative effectiveness of GA_3 on producing a similar response during auxin-induced xylogenesis in tissue cultures (Chap. 4.5). For example, the initiation of differentiation in bean callus was inhibited by the addition of GA_3 to the xylogenic medium (Haddon and Northcote 1976c). Investigators from three different laboratories found that exogenous GA_3 had a stimulatory effect on vascular differentiation in tuber explants of *Helianthus tuberosus* (Gautheret 1961 a, 1966, Dalessandro 1973, Saussay and Gautheret 1974, Phillips and Dodds 1977). Halperin and his colleagues found that GA_3 inhibited xylogenesis in the same plant material (Minocha and Halperin 1974, Watson and Halperin 1981, Minocha 1984). There are several possible explanations for these conflicting results. Autoclaving freshly prepared solutions of GA_3 can reduce the biological activity of the hormone by more than 90% (van Bragt and Pierik 1971). The weak xylogenic response reported by Minocha and Halperin (1974) could have resulted from their use of GA_3 in autoclaved media. This explanation appears unlikely, since Watson and Halperin (1981) found that filter-sterilized GA_3 also blocked xylogenesis in their cultures. Probably the inhibitory effect was related to their use of liquid cultures. Halperin's group found that GA_3 had relatively little effect on xylogenesis when tuber explants were cultured on an agar medium (Watson and Halperin 1981). Varietal differences in the content of endogenous hormones in the tubers may be a factor (Chap. 4.5). Unfortunately, primary explants taken from Jerusalem artichoke tubers have never been analyzed for endogenous hormones prior to an experiment on auxin-induced xylogenesis.

Endogenous gibberellins and gibberellin-like substances are present during experiments involving pith explants of *Lactuca* (Pearce et al. 1987). The major GAs identified in lettuce pith explants by gas chromatography-mass spectrometry (GC-MS) and GC-selected ion monitoring (GC-SIM) are GA_1, 3-epi-GA_1, GA_8, GA_{19}, and GA_{20}. During treatment of the explants under xylogenic conditions (IAA 57 μM, kinetin 0.5 μM) the concentration of an unknown biologically active polar GA-like substance(s) increased during the first 2 days of culture, although all of the identified GAs decreased as measured by GC-SIM. In explants cultured without exogenous hormones, or with IAA or kinetin alone, the concen-

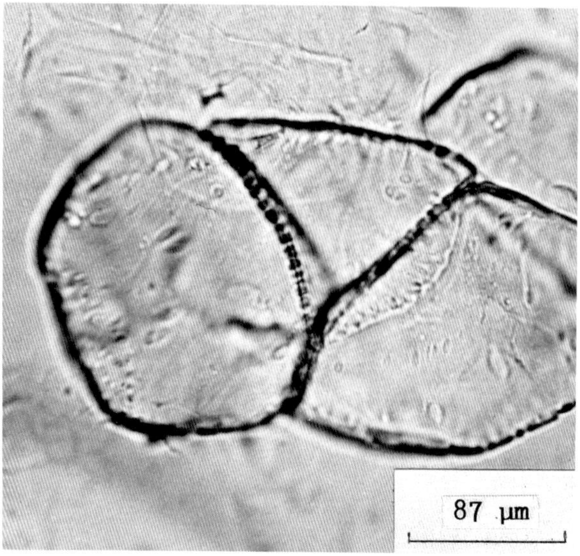

Fig. 2.2. Xylogenesis induced in the presence of gibberellic acid in explant of lettuce pith produce large numbers of parenchyma-like cells that stain with safranin 0. Although the cells are pitted, they do not contain secondary wall striations. (Roberts 1976)

tration of the polar GA-like compound(s) decreased, as well as the known GAs. The addition of GA_1 to the medium, in the absence of the other hormones, did not induce xylogenesis. GA_1 at very low concentrations (0.0014 µM, 0.003 µM), however, synergized auxin-induced xylogenesis in the lettuce pith system.

Xylogenic cultures in the presence of GA_3 frequently display a unique type of cytodifferentiation in the form of large parenchyma-like cells that stain weakly with safranin 0 (Roberts 1976, Fig. 2.2). Although the thin pitted walls presumably contain lignin, there is no evidence of secondary wall thickening characteristic of tracheary elements. In vitro experiments involving GA_3 indicate that these are anomalous xylem fibers (Aloni unpublished). These fiber-like cells have been observed in pith cultures of lettuce (Cawthon 1972), tobacco cultures (Snijman 1972), and in cultured hypocotyl segments of *Helianthus annuus* (Aloni 1982). The development of thin cell walls is also associated with the differentiation of *Coleus* phloem fibers in the presence of high ratios of GA_3: IAA (Aloni 1979). In tissue cultures high concentrations of GA_3 induced the differentiation of large numbers of xylem fibers coupled with the production of relatively few tracheary elements. The early stages of xylem fiber differentiation appeared to require a combination of auxin, cytokinin, and gibberellin (Aloni 1982). A hypothesis can be suggested that xylem fibers are formed instead of tracheary elements, because of a malfunction of microtubules in the patterned deposition of secondary wall thickenings. Several studies have provided evidence that GA_3 exerts some regulatory control on microtubule arrangement (Shibaoka 1974, Mita and Shibaoka 1984a, b, Mita and Katsumi 1986). The cellular orientation of microtubules may be related to the stability of this organelle to low temperature stress (Akashi and Shibaoka 1987, Chap. 5.2). Ultrastructural studies are needed on the arrangement of microtubules during xylogenesis induced in the presence of GA_3.

Some reports indicate that GA_3 enhances phloem differentiation. DeMaggio (1966) observed that cultured explants of *Pinus strobus* produced greater numbers of phloem elements in the presence of GA_3 than in the absence of the hormone. Combinations of auxin and GA_3 sprayed on seedlings of *Pinus brutia* stimulated phloem formation in the needles, whereas the application of auxin alone in lanolin was sufficient to increase phloem production in needles of *Pinus strobus* (Ewers and Aloni 1985). In both species the size and appearance of the cambial zone in the needles was unaffected by the hormone treatments, and evidently the rates of cell division and cytodifferentiation were approximately equal. The authors concluded that, in addition to endogenous auxin and GA_3, there is probably an unidentified tracheary differentiation factor produced in young, but not old, needles that limits tracheid differentiation in the mature needle. This supports the previous findings of Savidge (1983 b) (Savidge and Wareing 1981 a). Primary phloem fiber formation is strongly influenced by GA_3 when sprayed on intact plants (Stant 1961, 1963) and by combinations of auxin and GA_3 applied to defoliated stems (Aloni 1979, Chap. 4.2). Both phloem and xylem fiber differentiation are induced by spraying various plant species with combinations of auxin and GA_3 (US patent no. 4507144 issued to Aloni 1985).

2.5 Ethylene

Evidence linking ethylene to tracheary element differentiation is presented in Chap. 5.5 and 5.6. Ethylene promoted phloem differentiation in seedlings of *Ulmus* (Yamamoto et al. 1987 a), *Pinus* (Yamamoto and Kozlowski 1987 b), and *Thuja* (Yamamoto and Kozlowski 1987 c). The current data support the hypothesis that ethylene has a synergistic effect on xylogenesis in the presence of auxin and cytokinin. The role(s) of ethylene in xylem formation are unknown. One hypothesis is that ethylene controls lignification in differentiating xylem cells by the regulation of wall-bound peroxidase activity (A. R. Miller et al. 1985). Additional evidence concerning lignin-related pathways comes from observations that both polyphenolic biosynthesis during gummosis in *Prunus* and kino release in *Eucalyptus* are ethylene-mediated pathways (Nelson and Hillis 1978 a, Olien and Bukovac 1982).

2.6 Polyamines

Phillips et al. (1987) conducted a study of the role of polyamines on influencing the induction of cell division and xylogenesis in explants excised from dormant artichoke (*Helianthus tuberosus*) tuber. In the presence of exogenous auxin and cytokinin the sequential events include: an activation phase (0–24 h), mitotic activity (24–48 h), and xylem differentiation (after 48 h). Large numbers of tracheary elements formed in these cultures between 60 and 72 h. Endogenous spermidine and diaminopropane showed the greatest increases during the first 24 h of culture, and these polyamines then declined during the remainder of the experiment. There was no apparent relationship between polyamine synthesis and xylogenesis. The addition of methylglyoxal bis-guanylhydrazone (MGBG), an inhibitor of S-adenosyl-methionine decarboxylase (Holta et al. 1973), had no discernible effect on either cell division or spermidine accumulation in the cultures. Xylogenesis, however, was strongly inhibited by the presence of MGBG in the medium. Subsequent experiments by Phillips (personal communication) revealed that exogenous spermidine (>1 mM) completely inhibited the induction of xylogenesis in cultured explants of artichoke tuber (induction hormones: 2,4-D, BAP). Further experiments have not been performed to elucidate the inhibitory effect. One possibility involves the biosynthesis of xylogenic levels of ethylene. Spermidine decreased the endogenous level of ACC and inhibited the conversion of ACC to ethylene in soybean hypocotyl tissue (Suttle 1981). The observation that ethylene production from ACC is blocked by spermidine has been confirmed by others (H. Fuhrer et al. 1982, Shih et al. 1982).

The effects of exogenous ornithine, arginine, and polyamines on organogenesis in cultured leaf explants of *Datura innoxia* have been reported (Chriqui et al. 1986). Synergistic effects between ornithine and IAA on rhizogenesis were detected, whereas arginine was ineffective. A positive correlation was found to exist between rhizogenesis, ornithine decarboxylase activity (EC 4.1.1.17), and increases in endogenous putrescine and spermidine.

A preliminary investigation suggests that the biosynthesis of cadaverine is positively correlated with the induction of xylogenesis in cultured lettuce pith explants (A. R. Miller personal communication).

2.7 Inhibitors

It is difficult to establish that an endogenous compound is acting in an inhibitory manner to regulate a physiological event (Leopold and Kriedemann 1975, Milborrow 1984). One problem concerns the difficulty in distinguishing between secondary metabolites acting as chemical defense mechanisms against predators and compounds that are genuine regulators in the hormonal sense (Milborrow 1984). According to Milborrow, most of the experimental evidence supporting the view that mono- and di-phenols regulate IAA oxidase activity is inconclusive. Although water stress can lead to modifications in cytodifferentiation, these effects may not be due to the release of abscisic acid. Ethylene is rapidly synthesized under the same environmental conditions.

The endogenous inhibitor most frequently examined in connection with vascular differentiation is abscisic acid, and opinions are divided concerning its regulatory properties in cytodifferentiation. According to T. J. Wodzicki and A. B. Wodzicki (1980) there is a gradual accumulation of abscisic acid in the cambial zone and youngest phloem in the basal region of *Pinus silvestris* stems near the end of the growing season. Increases in abscisic acid may inhibit polar auxin transport and, as a result, delay the autolytic breakdown of protoplasts during the last stage of tracheid differentiation. A delay in cytodifferentiation extends the maturation phase and results in the formation of thick-walled tracheids characteristic of conifer latewood. The apical application of abscisic acid reduced the amplitude of morphogenic auxin transport according to T. J. Wodzicki et al. (1979). Jenkins (1974) found that abscisic acid injected into stems of *Pinus radiata* seedlings during earlywood formation produced a sharp reduction in the radial diameter of tracheids and decreased the mitotic activity of the vascular cambium. Other workers have presented an alternative viewpoint. Seasonal cambial activity and xylem differentiation were followed anatomically in trees of *Pinus contorta* and, at the same time, the levels of IAA and abscisic acid were monitored within the tissues (Savidge and Wareing 1984). The concentration of abscisic acid showed relatively little variation throughout the growing season, and the level did not increase in the cambial region during the period of reduced radial expansion of differentiating cambial derivatives. There was no significant increase in abscisic acid as shoots, needles, and the cambium ceased growth. Cambial IAA levels remained high after shoot and needle extension had ceased, and the highest levels of IAA were located in the main stem near the crown base. The level of IAA decreased after cambial inactivation in the fall during latewood formation. Little and Wareing (1981) examined the seasonal fluctuations of IAA and abscisic acid in the cambial region of *Picea sitchensis*, and their evidence does not support a role for abscisic acid in the regulation of cambial activity.

Abscisic acid is capable of producing morphogenetic effects in tissue cultures. Bean callus was induced to form roots and vascular nodules by lowering the ratio of auxin to cytokinin in the medium. Ratios of auxin:cytokinin higher than 5:1 caused differentiation to cease. Both types of differentiation were completely inhibited by the addition of abscisic acid to the medium. Abscisic acid also inhibited the induction of both PAL and β-1,3-glucan synthetase activities (Haddon and Northcote 1976c). Although abscisic acid suppressed xylogenesis in cultured tuber explants of *Helianthus tuberosus*, cell division was stimulated (Minocha and Halperin 1974, Minocha 1979, Minocha and DiBona 1979). Additional studies by Watson and Halperin (1981) revealed that the inhibition of xylogenesis varied from 25% up to 90% of the control cell counts (see Minocha 1984). Abscisic acid is known to regulate foliar morphogenesis in aquatic plants (Anderson 1978, B.L.L. Liu 1984). Bornman (1983) reviewed the influences of abscisic acid on morphogenesis and plant anatomy. He considers that abscisic acid may be involved in the enhancement of proline production, and accumulation of this amino acid may influence tracheary element differentiation (Roberts and Baba 1968). Although the modes of action of abscisic acid are not definitely established, the collective evidence suggests that abscisic acid could have an effect on morphological development by either (1) interacting with IAA, gibberellins, cytokinins, and ethylene; (2) influencing RNA and protein synthesis; (3) altering membrane characteristics; or (4) stimulating the production of some stress-induced compound such as proline (Bornman 1983).

Davies (1987) has suggested in a preliminary report that ABA, synthesized in root tips during periods of water stress, may be transported to the shoot system via the transpiration stream. It is conceivable that this mobile ABA within the xylem may influence vascular differentiation.

2.8 Summary

The initiation of vascular differentiation occurs in progenitor cells as a result of some primary event occurring in the presence of auxin, cytokinin, and ethylene.

Although auxin is necessary for initiation of the cytodifferentiation process, it may or may not be directly involved in the primary event. The regulatory roles of the other two hormones cannot be ruled out, and there is increasing evidence of the involvement of second messengers. Aside from a synergistic effect on fiber development in the presence of auxin, the role of gibberellin in vascular differentiation is not clearly established. Additional studies must be conducted with gibberellins, and, in particular, with the possible roles of gibberellin-like compounds. The evidence to date is inconclusive that abscisic acid or any other naturally occurring substances of an inhibitory nature play a hormonal role in the differentiation of xylem and phloem elements.

Chapter 3

Vascular Differentiation Within the Plant

R. ALONI

3.1 Major Problems

3.1.1 In Vivo Versus In Vitro Systems

The problems dealt with in this chapter concern the control of organized or patterned differentiation of vascular tissues within the plant body by growth regulators. This organized differentiation of vascular tissues is characterized by a high degree of order. The tissues are formed in a predictable three-dimensional structure, which is typical of each species. Vascular tissues are formed continuously in the intact plant, as long as the plant grows from apical and lateral meristems. New tissues develop in dynamic relationship to one another. The vascular system is complex. It is composed of two kinds of conducting tissues: the phloem and the xylem, each of which is comprised of several types of cells (Fahn 1982).

There is evidence that the complexity of the tissues is controlled by a series of signals and is dependent upon interactions between the cells. Some of these control mechanisms will be discussed in this chapter. Vascular differentiation within the plant depends on shoot-to-root flux of signals, one of which is the polar flow of auxin from the young leaves toward the root tips. The polar movement of auxin through the cells induces their longitudinal polarity. The movement of auxin and other growth regulators along the meristems and the differentiating cells controls the well-organized patterns of different types of conducting and supporting elements within the vascular tissues. As the vascular system is formed along the pathways of movement of growth regulators, it results in a continuity of vascular tissues which extend along the entire plant and connect the leaves and other stem appendages with the roots (Sachs 1981, Aloni 1987b).

Tissue culture techniques are usually used for the study of different aspects of vascular differentiation (Chap. 4). Vascular differentiation occurring under tissue culture conditions, either in a callus grown on a solid medium or in a cell suspension grown in liquid culture, typically lacks consistent orientation. Vascular differentiation in tissue culture occurs at any angle and in random patterns. Isolated vascular elements, such as tracheids, are formed instead of longitudinal vessels along the plant axis. Callus tissues usually contain vascular nodules which are haphazard in their arrangement and orientation. Some degree of order can be found, such as the differentiation of xylem, which is usually oriented toward the center of the vascular nodule, whereas phloem is oriented toward the periphery and usually surrounds the xylem (Aloni 1980). The polarity of the cells, a basic feature of organized tissues and organized differentiation, is disrupted or does not exist in cultures. The same is true for the contacts and interrelationships between the cells.

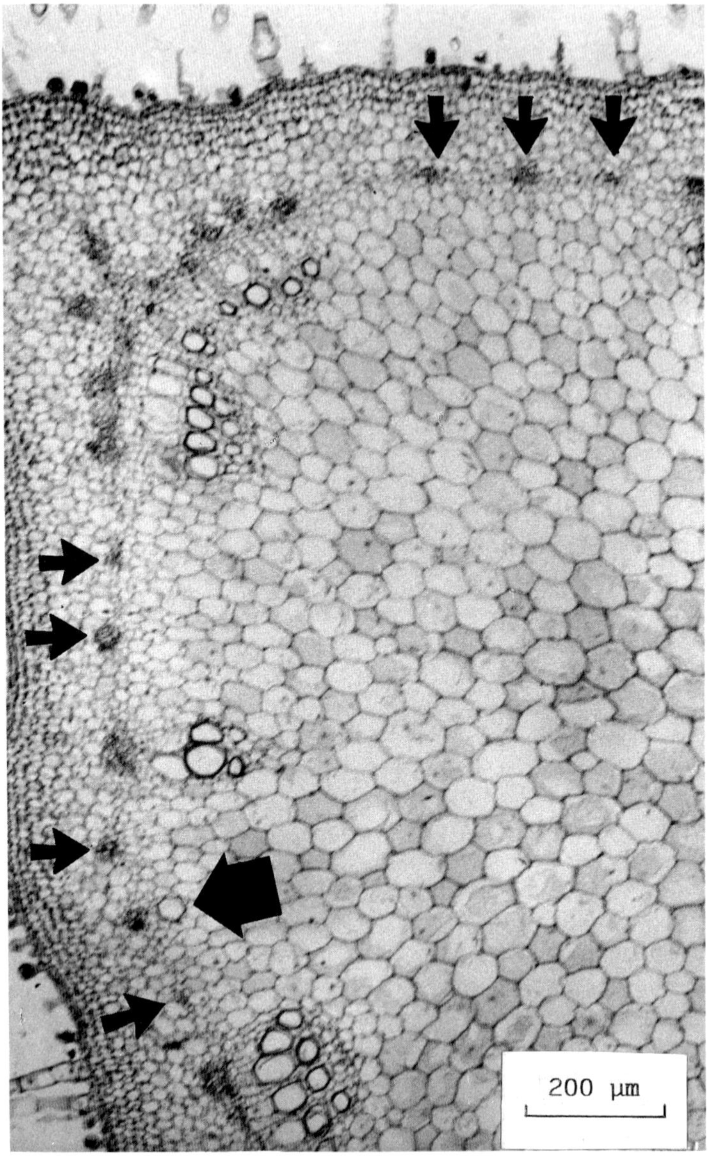

Fig. 3.1. Transverse section from the middle of an intact young internode of *Coleus blumei* showing the pattern of primary differentiation of phloem and xylem. Seven strands of phloem with no xylem are shown (*smaller arrows*). A new vessel, in an early stage of differentiation, is forming in a preexisting strand of phloem (*large arrows*). ×90. (R. Aloni unpublished)

Despite the fact that in vitro systems are useful experimental tools to answer questions of vascular differentiation at the biochemical, subcellular, and cellular levels, they always leave open questions about the validity of the findings obtained in these systems when organized differentiation of vascular tissues within the intact plant is concerned. We can expect to find modifications, or even different mechanisms in the differentiation of vascular tissues at the level of cell, tissue, and organ when in vitro and in vivo systems are compared.

3.1.2 The Difficulty of Observing the Phloem

In the intact plant, xylem does not differentiate in the absence of phloem, although phloem often develops in the absence of xylem (Aloni 1987 b). In addition, the phloem always develops before the xylem (Esau 1965 b). Therefore, any new xylem differentiation will follow the pattern of phloem differentiation. This basic pattern, both in space and time, can be easily observed in young internodes at early stages of vascular development (Fig. 3.1). It emphasizes the importance of understanding phloem differentiation as a basic background for studies of xylem differentiation. However, a student of vascular differentiation is faced with a major problem, that is, the difficulty of detecting sieve elements. In comparison with tracheary elements, easily recognized by their secondary wall thickenings, sieve elements usually have no visible markers except callose deposits associated with sieve plates. Therefore, it is not surprising that reports on phloem differentiation are confusing and often contradictory. The difficulty of locating sieve elements also explains the reason why most studies on vascular differentiation were restricted to the xylem. Clearing techniques have been developed for observing the three-dimensional structure of the phloem (LaMotte and W.P. Jacobs 1962, Bisalputra and Esau 1964, Aloni and Sachs 1973, C.A. Peterson and Fletcher 1973, Robbertse and McCully 1979, Lersten 1986, Prazak and C.A. Peterson 1987).

3.1.3 Primary Phloem Structure and the Zero-Day Control

The major difference between the three-dimensional structure of primary phloem systems and that of primary xylem systems, is that the phloem produced under summer conditions usually has a network pattern with many phloem anastomoses between the longitudinal vascular strands (Fig. 3.2). Students of vascular differentiation and phloem transport should remember that phloem anastomoses between the vascular strands are common in many plant species (Aloni and Sachs 1973). Their network differentiation depends on the growing conditions (Aloni and Sachs 1973, Aloni et al. 1986). The average number of phloem anastomoses in a summer-grown mature internode of *Coleus blumei* was 450 and 3000 in a similar internode of *Dahlia pinnata* (Aloni and Sachs 1973). These phloem anastomoses, which consist of one or more lateral sieve tubes, appear to function as rapid channels for lateral transport across stems and for movement of organic materials upwards from leaves.

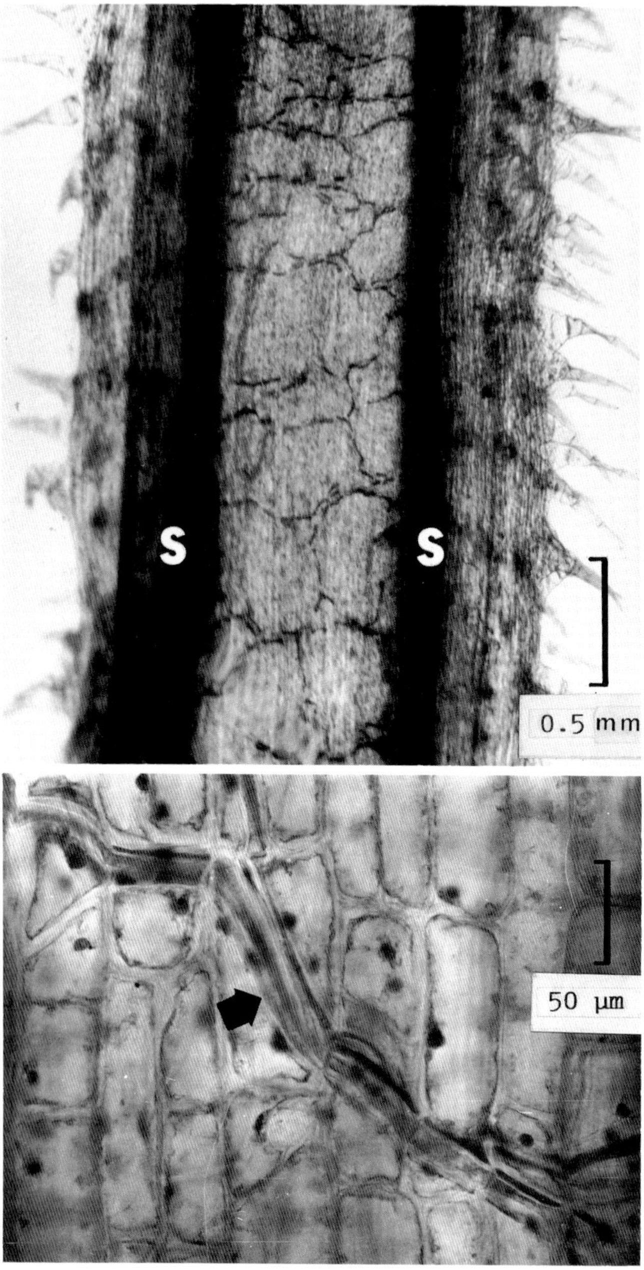

Fig. 3.2. Thick preparations, cleared with lactic acid and stained with lacmoid, showing phloem anastomoses. *Upper* Pattern of phloem anastomoses between two vertical strands (*S*) in hypocotyl of *Cucumis sativa.* ×35. *Lower* Phloem anastomosis (*arrow*) in mature internode of *Coleus blumei.* ×300. (Aloni and Sachs 1973)

Aloni and W. P. Jacobs (1977 b), during a study of phloem regeneration in the Princeton clone of *Coleus blumei*, found well-developed phloem anastomoses in zero-day control plants. These mature sieve tubes, between the longitudinal strands, were easily recognized since more callose was deposited on their sieve plates compared to callose associated with the newly formed sieve plates of the regenerated sieve elements. LaMotte and W. P. Jacobs (1962) explained which types of lateral sieve tubes between the longitudinal strands were not counted as regenerative phloem, in their example for a "sampled regeneration area." The use of a "sampled regeneration area" for the phloem, a term which was never needed for xylem studies, hints that there was a problem in deciding at what distance from the wound to stop searching for lateral sieve tubes. In published photographs of the "sampled regeneration area" (Thompson and W. P. Jacobs 1966, Houck and LaMotte 1977), one can clearly see phloem anastomoses, as they either pass through the wound or have no connection whatsoever to the wound site. Unfortunately, some of these phloem anastomoses were counted as regenerated sieve tubes.

Almost all of the studies of phloem regeneration around a wound lack the zero-day control, which reveals the number and density of phloem anastomoses in the intact internodes. Ignoring the existence of phloem anastomoses raises questions about the validity of the data reported. For example, Thompson (1967) reported a very fast phloem regeneration of 1.5 days following wounding of a mature internode of *Coleus*. Since phloem anastomoses may have been referred to as fast sieve tube regeneration, Aloni and W. P. Jacobs (1977 b) repeated experiments with the same *Coleus* clone under the same conditions. Sieve tube regeneration had not occurred in any of the plants after 2 days, and, to remove any doubt concerning the observations, two different clearing techniques were employed (LaMotte and W. P. Jacobs 1962, Aloni and Sachs 1973). The first regenerated sieve tubes were detected around the *Coleus* wounds only after 3 days (Aloni and W. P. Jacobs 1977 b).

The aim of this chapter is to review our present understanding of the roles of plant growth regulators in controlling the organized differentiation of phloem and xylem within the plant. It will focus on the role of auxin as it seems to be the major signal involved in the control of all apsects of organized differentiation of vascular tissues. The discussion will also focus on the roles of cytokinin and gibberellin. In addition, this chapter will attempt to show some basic principles of organized vascular differentiation in relation to the conduction and the adaptation of vascular systems.

3.2 Control of Vascular Differentiation by Auxin Flow

Developing buds and young growing leaves induce vascular differentiation, which extends from the developing leaves towards the root tips. Along the plant the vascular tissues are organized in longitudinal strands or in a continuous cylinder, in predictable patterns for each plant species. Auxin produced by the young leaves was found to be the limiting and controlling factor for both phloem and xylem differentiation (W. P. Jacobs 1952, LaMotte and W. P. Jacobs 1963,

Thompson and W. P. Jacobs 1966, Bruck and Paolillo 1984 e, Aloni 1987 b). Two major IAA transport pathways are recognized along the plant axis (Morris and Kadir 1972). The first pathway involves the polar flow of auxin which induces vascular differentiation. This pathway originates in the young leaves and moves toward the root tips via procambium, cambium, differentiating vascular elements and parenchyma cells (Wangermann 1974, Morris and Thomas 1978). This pathway is sensitive to auxin transport inhibitors (Morris et al. 1973, Meicenheimer and Larson 1985). Evidence indicates that the polar flow of auxin moves in waves, that is, by oscillations of the transport rate (Hertel and Flory 1968, Zajaczkowski and T. J. Wodzicki 1978, T. J. Wodzicki et al. 1979, 1984, Zakrzewski 1983, Zajaczkowski et al. 1984). A signal which moves along the plant axis in a wave form may also convey morphogenetic and positional information (Goodwin and Cohen 1969). This concept merits further investigation. The second IAA transport pathway involves the rapid nonpolar movement of the hormone through sieve tubes. The auxin which moves in the phloem originates in mature leaves (Bonnemain 1971, Morris et al. 1973, Goldsmith et al. 1974). The nonpolar movement of auxin via the sieve tubes may have a promoting effect on vascular regeneration in cases of sieve tube rupture below the mature leaves. The promoting effect is probably a result of additional auxin in the wound region that arrives from the wounded phloem (Benayoun et al. 1975).

Developmental studies on the differentiation of vascular tissues in the stems of different plants have shown a clear relationship between the patterns of the leaves and the patterns of vascular tissues descending from the leaves toward the roots. The development of the vascular tissues is closely correlated with leaf development (Esau 1965 b). A series of experiments, mainly performed on pea seedlings, led Sachs (1969, 1981, 1986) to the hypothesis that canalization of auxin flux determines the orderly pattern of vascular differentiation. According to this hypothesis, auxin movement from a young leaf occurs initially by diffusion. As the auxin diffuses through cells it induces the phenomenon of polar auxin transport. The relation between transport and differentiation is one of a positive feedback: the more auxin flows, the better the cells transport auxin. In other words, differentiation increases the ability of the cells to transport the same signal which induces differentiation. This leads to canalization of further auxin flow along a narrow file of cells. Horizontal diffusion of auxin around a wound induces the cells to transport auxin along a new horizontal polarity (Sachs 1975, Gersani and Sachs 1984, Gersani 1987 b). The continuous polar flow of auxin induces a further complex series of events which finally leads to the formation of a defined vascular strand. The vascular strand remains the preferential pathway for the continuous polar transport of auxin, as cells possessing the ability to transport auxin are associated with vascular tissues (Morris and Thomas 1978, M. Jacobs and Gilbert 1983, M. Jacobs and Short 1986). As the vascular strands are the fastest pathway for auxin movement (Lachaud and Bonnemain 1984), these strands attract new streams of auxin from the young developing leaves and thus a continuous network of vascular tissue is formed. Contact between new and preexisting vascular strands is promoted when the auxin flux in a preexisting strand is low (for example, one descending from an old leaf), and contact is prevented as long as the strand is well supplied with auxin (for example, a strand descending from a young leaf). Experiments with the stem anatomy of *Coleus* support Sachs' (1969) concept that auxin relations determine the three-dimensional patterns of xylem strands in the intact plant (Bruck and Paolillo 1984c, d).

3.3 Role of Cytokinin in Vascular Differentiation

Roots do not induce vascular differentiation. Roots need not be present in order to obtain vascular differentiation in stem tissue (Thompson and W.P. Jacobs 1966). Roots, however, affect vascular differentiation by orienting the pattern of vascular differentiation from the leaves towards the root tips by acting as a sink for the polar flow of auxin originating in the young leaves (Sachs 1968).

The root apices are sources of inductive stimuli that promote vascular development. The major stimulus is cytokinin. Most of our information concerning the role of cytokinin in vascular differentiation comes from in vitro studies (Chap. 2.4). Very little information has been gained about the role of cytokinin from studies done with intact plants. Therefore, the role of cytokinin in organized differentiation of vascular tissues within the plant body is poorly understood. From tissue culture experiments cytokinin is known to stimulate cell division and to control both the differentiation of tracheary elements (Dalessandro and Roberts 1971, Dalessandro 1973) and secondary xylem fibers (Aloni 1982). Studies are needed with intact plants in order to confirm the findings obtained with cytokinin in culture. Torrey et al. (1971) suggested that the difficulty in proving involvement of cytokinin in the control of vascular differentiation in the intact plant results from the relatively high levels of root-supplied cytokinin within the tissues. The difficulty in studying the role of cytokinin in vascular differentiation, as well as other processes in the plant, is further increased by the lack of anticytokinins and by the fact that excised organs and tissues are capable of synthesizing cytokinin under experimental conditions.

Houck and LaMotte (1977) reported that zeatin or zeatin riboside in aqueous solution, applied to the bases of excised internodes of *Coleus* receiving IAA at their apical ends, restored phloem regeneration to the level of that found in whole plants. However, other cytokinins (kinetin, kinetin riboside, 2iP, 2iPA) failed to promote phloem regeneration in the excised internodes of *Coleus*.

The role of cytokinin was studied during fiber differentiation in intact *Helianthus annuus* plants (Saks et al. 1984). Reducing the endogenous cytokinin supply, either by side root removal or by lowering the rate of transpiration, decreased the formation of secondary xylem fibers in the hypocotyl of the young *Helianthus* plants. This decrease was considerably reversed by the addition of kinetin to the roots. The application of kinetin to intact plants via the roots promoted the differentiation of fibers in the hypocotyl (Fig. 3.3). The addition of kinetin to the roots had a moderate promoting influence on fiber differentiation within the strands and a very pronounced effect on fiber differentiation between the strands. A brief exposure of intact *Helianthus* plants to kinetin, when the reagent was applied to the root of the young plants, exerted a promoting effect on fiber differentiation in the stem. This effect was detected only a few days after the removal of kinetin from the medium. This delayed promoting effect can be interpreted as

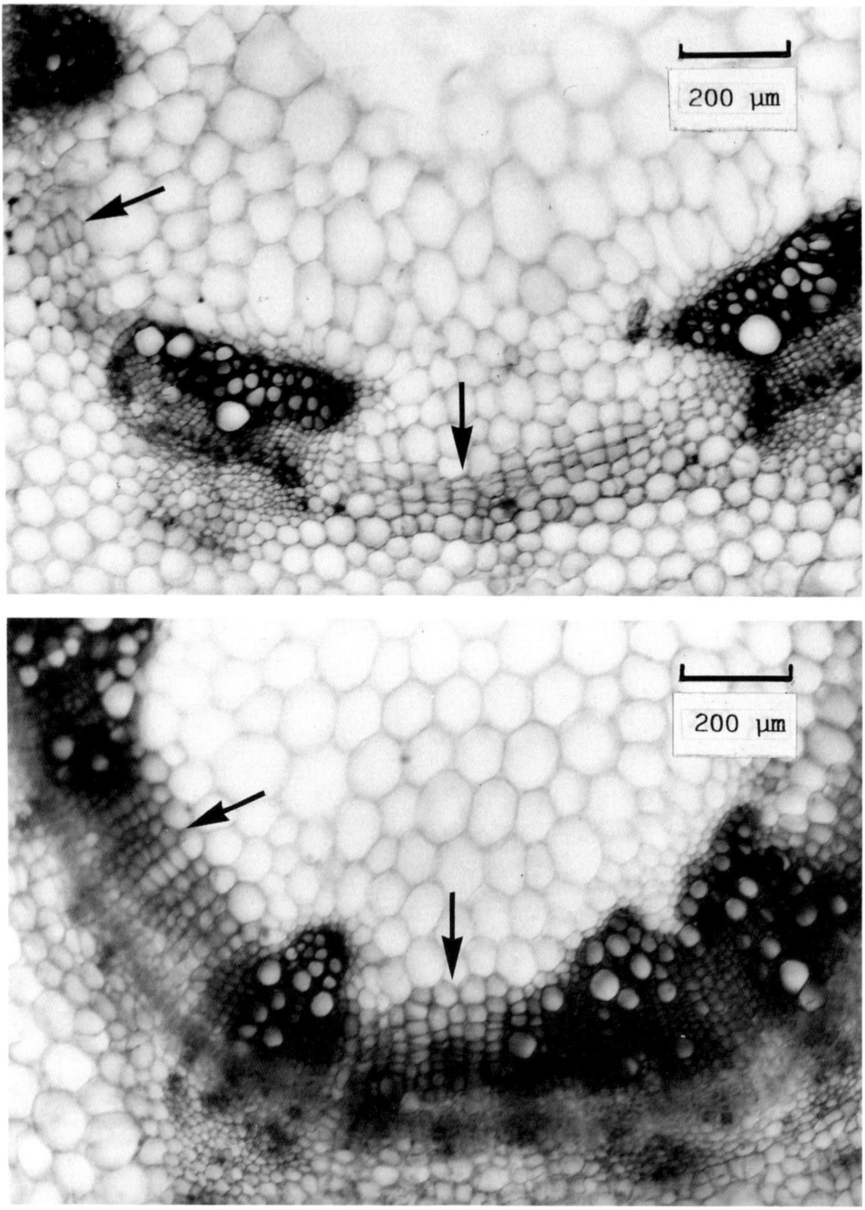

Fig. 3.3. Transverse sections from middle of the hypocotyl of a young *Helianthus annuus* plant showing the pattern of secondary xylem fibers induced by kinetin following application to the roots of intact plants for a period of 21 days. Both photomicrographs were taken from the same experiment and with the same magnification. ×80. *Upper* Intact control shows the typical strands with interfascicular cambium (*arrows*) between them after 21 days of growth in Hoagland's medium without kinetin. *Lower* Treatment with kinetin (0.25 μg^{-ml} in Hoagland's medium) induced secondary xylem fiber differentiation between the strands (*arrows*). Note there are more fibers within the strands in the kinetin-treated plant than in the control plant. (Saks et al. 1984)

the influence of kinetin on early stages of fiber differentiation, probably affecting the rate of cell division in the vascular cambium (Saks et al. 1984).

3.4 Relation Between Phloem and Xylem Differentiation

The relations of phloem and xylem in the intact plant must form the background to the understanding of the role of plant growth regulators in the control of these two complex tissues. Therefore, let us look for their relations in the different parts of the intact plant. In the stem, the xylem is formed on the inner side toward the pith, and the phloem toward the outside. In some plant families phloem is found both at the outer and at the inner side of the xylem. Xylem does not differentiate in the absence of phloem formation within the intact plant. Phloem often differentiates, however, in the absence of xylem formation (Aloni 1987 b).

In the embryo and in the young organs, the primary phloem differentiates before the primary xylem (Esau 1965 b). In the stem of angiosperms, in addition to strands which consist of xylem and phloem (collateral strands), there are strands of phloem only (Aloni and W. P. Jacobs 1977 b). For example, in the young internodes of *Coleus* there are more strands with only phloem than collateral strands (Aloni 1987 a). In mature internodes of *Coleus*, on each collateral strand there is usually a strand of phloem with no xylem (Aloni and W. P. Jacobs 1977 b, Bruck and Paolillo 1984 e). Based on experiments with IAA, Bruck and Paolillo (1984 e) regarded the phloem-only strands as an early auxin-deficient stage of collateral strand formation, which depends on auxin availability to the strand for fulfillment. In the stem of many plant species there are phloem anastomoses (lateral sieve tubes with no xylem) that occur between the longitudinal strands (Fig. 3.2), and these structures facilitate lateral movement of assimilates (Aloni and Sachs 1973). In cucumber seedlings the number and density of phloem anastomoses are dependent on light intensity (Aloni et al. 1986). Aloni (1987 a) proposed that phloem anastomoses are induced by low level streams of auxin. This view is supported by the fact that when high auxin concentrations were applied to decapitated young stems of *Luffa* this resulted in the differentiation of anastomoses of phloem with xylem (Aloni 1987 a).

A network of sieve tubes with no xylem was found in the fleshy region of the fruit of *Coccinia grandis* with no connections to the main strands of the fruit (Shah et al. 1983). Similarly, a mycelium-like network of internal phloem with no xylem occurred in the inner mesocarp of the lateral pod walls of the fruit of *Vigna unguiculata* (Kuo and Pate 1985). These sieve tubes were probably induced by low levels of auxin.

In mature organs, the phloem and xylem tissues are produced by the cambium and their realtions are different. In mature needle leaves of *Pinus*, secondary phloem, but not secondary xylem, is produced each year (Ewers 1982 a, b, Ewers and Aloni 1985). When the needles are young, however, both xylem and phloem are formed. In 1–33 year-old needles of *Pinus longaeva* the vascular cambium remained active throughout the life of the needles and produced secondary phloem unidirectionally (Ewers 1982 a, b). In needle leaves of *Pinus brutia*, grown in Israel, the secondary phloem is produced throughout the year except in midsummer. This is contrasted by an autumn and winter dormancy in secondary phloem formation in needles of *Pinus strobus* grown in Massachusetts, USA. In each country the dormancy period in secondary phloem production occurred in a locally difficult season: in the arid summer in Israel and in the cold winter in Massachusetts (Ewers and Aloni 1987).

A completely different relationship occurs between the secondary vascular tissues in the stem of trees. Although the secondary phloem may precede secondary xylem by several weeks (Evert 1963), the number of cells produced by the cambium during the season toward the xylem is usually greater than that produced toward the phloem. The ratio of xylem : phloem cells in *Eucalyptus camaldulensis* was 4 : 1 (Waisel et al. 1966), in *Carya pecan* 5 : 1 (Artschwager 1950), and 6 : 1 in *Cupressus sempervirens* (Liphschitz et al. 1981). A higher ratio of xylem : phloem (15 : 1) was reported for fast-growing branches of *Thuja occidentalis* (Bannan 1955) as well as in *Abies concolor* (14 : 1) (Wilson 1963).

Quantitative changes that occur between the phloem and the xylem in the primary and secondary tissues, as well as in the secondary tissues in needles versus stem tissues, are probably due to differences in levels and ratios of growth regulators flowing through the procambium, cambium, and differentiating vascular elements. A possible role of sugar, as a factor controlling the relation between phloem and xylem, can be eliminated (Aloni 1987b). Both auxin and gibberellin may be involved in controlling the relationship between phloem and xylem differentiation (Wareing et al. 1964, Digby and Wareing 1966, Lachaud 1983, Zakrzewski 1983, Lachaud and Bonnemain 1984, Ewers and Aloni 1985 McKee and Morris 1986). The relative levels of applied auxin and gibberellin were important in determining whether or not mainly xylem or phloem tissue was produced in stems of *Populus robusta* (Digby and Wareing 1966). High IAA/low GA_3 concentrations favored xylem formation, whereas low IAA/high GA_3 levels favored phloem differentiation. The picture is far from being complete, however, as there are some contradictory results which might have arisen from the use of different experimental systems. For example, gibberellin promoted phloem differentiation in leaves (Ewers and Aloni 1985) and stem tissues (DeMaggio 1966), but reduced phloem production and decreased the phloem : xylem ratio in the storage root of carrot (McKee and Morris 1986). Difference in sensitivity of the meristems to the stimuli may also affect the ratio of xylem and phloem production in young and old organs.

3.5 Graniferous Tracheary Elements

Special vascular elements, which were first termed "phloeotracheids", are found in the haustoria of certain root parasites. This term was coined by Benson (1910), since she believed that they combined the functions of phloem and xylem. About 10 years ago, Fineran et al. (1978) found that these cells are tracheids or vessel elements and have no sieve element features (Fig. 3.4). The cells were therefore renamed graniferous tracheary elements. These elements are characterized by granular, amorphous, or fibrillar materials which are included in their lumen (Fineran 1985). Tubule substructure and crystalloids were also found in the lumen of graniferous tracheary elements (Fineran et al. 1987).

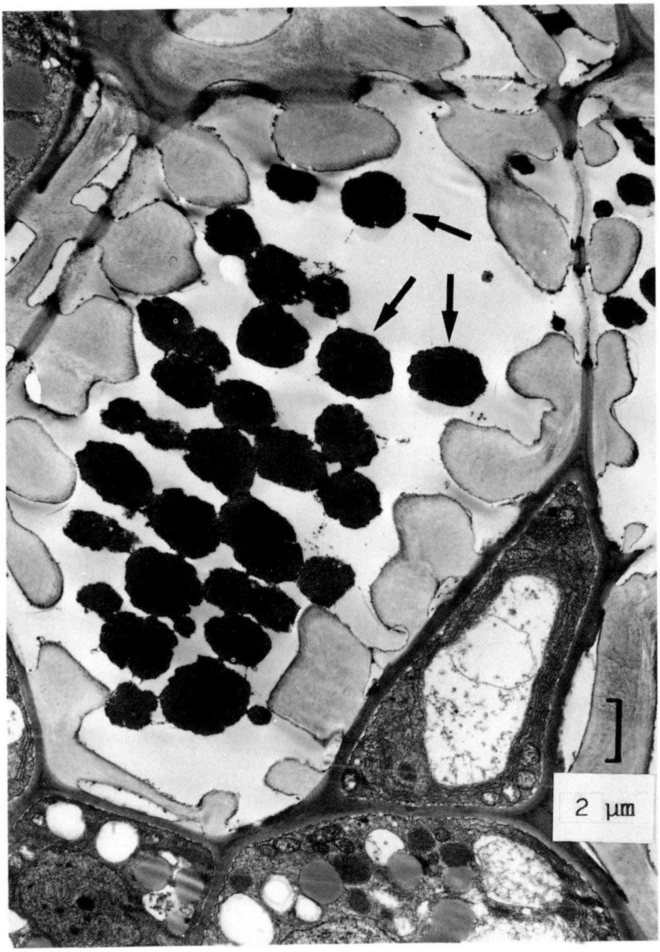

Fig. 3.4. Graniferous tracheary element in the haustorium of *Exocarpus bidwillii*. Note the homogeneous composition of the granule matrix (*arrows*). ×4000. (Transmission electron micrograph courtesy of B. A. Fineran)

The included materials may be proteins (Fineran 1983, Niranjana and Shivamurthy 1987) derived during the onotogeny of the cell from endoplasmic reticulum or starch originating from amyloplasts (Dobbins and Kuijt 1973). Granules rich in starch, lignin, and tannin were also reported (Weber and Hildenbrand 1978).

The role, if any, of graniferous tracheary elements in regulating ascent of sap from host to parasite is still an open question. This role has not been investigated under experimental conditions. Fineran and Ingerfeld (1982) suggested that graniferous tracheary elements may reduce the risk of back-flow of xylem sap from parasite to host under conditions of water stress in the host. This suggested physiological function might take place in elements containing materials composed of fibrillose or amorphous masses, which can serve as a plugging matrix. However, the same graniferous tracheary elements may reduce the ascent of sap from the host to the parasite as well.

The control of graniferous tracheary element differentiation has not yet been investigated experimentally. It is reasonable to assume that these xylem elements are induced and controlled by the polar flow of auxin originating in the young leaves of the parasite. Cytokinin originating either in the haustorium, or in the host plant, may also be involved.

3.6 Control of Conduit Size and Density

Generally, there is a continuous increase in the size of individual vascular elements from leaves to roots. This increase is observed in both tracheid length and diameter as one proceeds from branches to trunk and down into the roots of *Sequoia sempervirens* trees (Bailey 1958). Similarly, a gradual increase was found in vessel diameter and vessel length from twigs, down along the shoot, and extending into the roots of *Acer rubrum* trees (Zimmermann and Potter 1982). The basipetal increase in vessel diameter is associated with a decrease in vessel density, that is, the number of vessels per unit of transverse-sectional area.

The increase in conduit size can also be observed in a transverse section in a stem or a branch of a tree. The increase proceeds outward from the inner growth ring, through a number of annual growth rings, until a constant size is attained. In extreme old age (400–500 years), a decrease in conduit size was observed (Dinwoodie 1961). However, in the oldest stems of *Pinus longaeva*, tracheid length has steadily increased over the last 2200 years, with no signs of leveling off (Baas et al. 1986).

A hypothesis explaining the axial changes that occur in vascular element size and density from leaves to roots was suggested recently (Aloni and Zimmermann 1983). Based on the polar flow of auxin as the major factor controlling vessel differentiation, the following six-point hypothesis was developed: (1) Basipetal polar transport of auxin from leaves to roots results in a gradient of decreasing auxin concentration progressively down the axis of the plant. (2) Structural or physiological obstructions to auxin flow result in localized increases in auxin concentration. (3) The distance from the auxin source to the differentiating vessel element will determine the amount of auxin the progenitor cell will receive. (4) The rate of vessel differentiation is determined by the amount of auxin the differentiating cell receives: a high concentration results in rapid differentiation, whereas low concentrations cause slow development. Therefore, the duration of the differentiation process increases from shoots to roots. (5) The final diameter of a vessel is determined by the rate of the differentiation process, since cell enlargement obviously ceases at the time of secondary wall deposition. Rapid differentiation results in narrow vessels, whereas slow differentiation permits more cell expansion and thus larger vessels. Decreasing auxin concentration from leaves to roots results in a corresponding increase in vessel diameter in the same direction. (6) Vessel density is also regulated by auxin concentration. The higher the concentration of auxin, the higher the density. The density therefore decreases from shoots to roots.

Although there is evidence that a gradient of decreasing auxin content from leaves to roots occurs along the main stem of *Cucumis sativus* plants (point 1) (Galun et al. 1965), there still remains much work to be done to confirm this in other species.

The first test of the hypothesis was made by examining vessel differentiation in the secondary xylem of *Phaseolus vulgaris* seedlings on the basis of applied auxin concentration and axial distance from the point of application. NAA (0.03, 0.1, 1.0% w/w) was applied in lanolin to decapitated bean seedlings. The rate of secondary xylem differentiation decreased basipetally with an increase in distance from the auxin source (point 4). There was a basipetal gradient in vessel diameter associated with the distance from the auxin source and with auxin concentration, that is, the highest auxin concentration produced the narrowest vessels (point 5) (Figs. 3.5, 3.6). Vessel density decreased basipetally with increasing distance from the site of NAA application, an observation in agreement with point 6 of the hypothesis (Fig. 3.6). A subsequent study was made of the second point of the hypothesis, that is, the effect of obstructing auxin flow on vessel differentiation (Aloni and Zimmermann 1984). This was accomplished by a partial stem girdle in *Acer rubrum*, an operation that presumably raised the auxin concentration locally. The regenerated vessels in the restricted area were small in diameter with an increase in vessel density, a finding in agreement with the hypothesis. Similarly, narrow regenerative vessels differentiated in the first formed xylem that developed in girdled stems of lianas (Dobbins and Fisher 1986).

The six-point hypothesis (Aloni and Zimmermann 1983) is also applicable to explain axial changes in the diameter and length of fibers (Denne and Whitbread 1978, Aloni and Gad 1982, Gad and Aloni 1984).

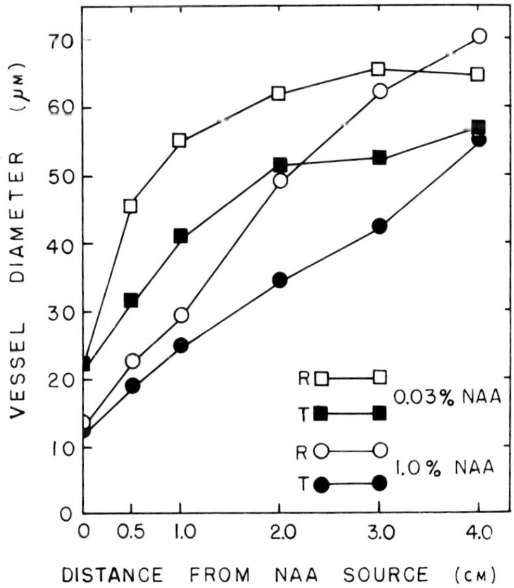

Fig. 3.5. Effects of low (0.03% NAA, *squares*) and high (1.0% NAA, *circles*) auxin concentrations on the radial (*R*) and tangential (*T*) diameters of late-formed vessels produced along decapitated internodes of *Phaseolus vulgaris*. Both concentrations induced a substantial gradient of increasing vessel diameter with increasing distance from the site of application. The high concentration yielded the narrowest vessels immediately below the site of application. (Aloni and Zimmermann 1983)

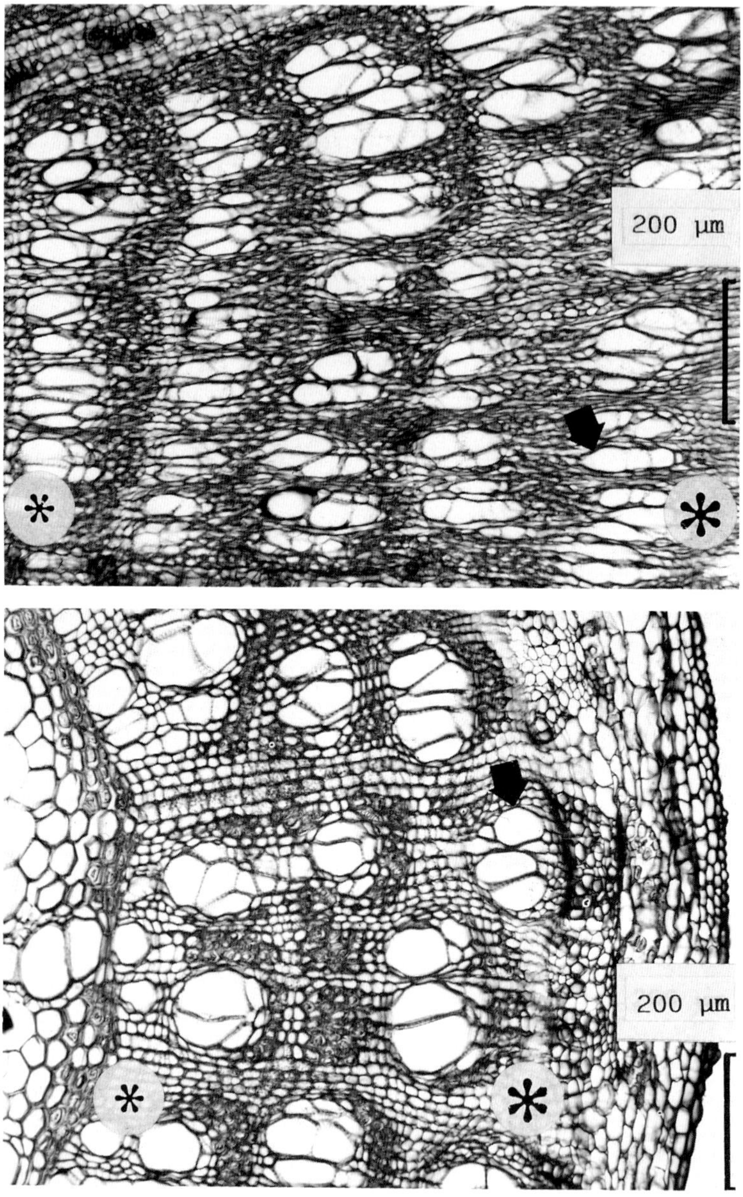

Fig. 3.6. Transverse sections prepared from the same internode of *Phaseolus vulgaris* following treatment with NAA (0.1% w/w) in lanolin. Both sections are oriented in the same direction and are of the same magnification. ×100. *Upper* Section taken 5 mm below site of application. *Lower* Section taken 40 mm below site of application. Note the increase in vessel diameter and decrease in vessel density with increasing distance from the auxin source. Changes of vessel pattern from layers (*upper*) to strands (*lower*) are also evident. More secondary xylem was induced by auxin at 5 mm (*upper*) than at 40 mm (*lower*). These changes in secondary xylem width, marked by *stars*, show equivalent locations in the photomicrographs. The *stars* mark the borders of the secondary xylem areas that were induced by the auxin during the 20-day experimental period. The *arrow* marks a late-formed secondary vessel. (Aloni and Zimmermann 1983)

3.7 Sites of Xylem Initiation and Rates of Maturation in Relation to Conduction

When the differentiation of the vascular system of the entire plant is considered, the nodes and junctions of organs are of particular interest both from a developmental and a functional viewpoint. The second point of the six-point hypothesis (Aloni and Zimmermann 1983, 1984) suggests that a local structural or physiological obstruction of the flow of auxin, for example, at nodes and junctions where two flows of auxin merge, will raise the auxin concentration locally. This local increase in auxin concentration is expected to result in fast differentiation of numerous small vascular elements. This assumption is in accordance with the fact that the first tracheary elements are initiated in the nodes at the bases of leaf primordia. The fast differentiation in the nodes results in a discontinuous pattern of xylem initiation. From each node the xylem differentiates bidirectionally until the isolated loci of xylem become connected. This general pattern of early xylem development was found in the young portions of the stems of a variety of plants (Esau 1965 b). In *Coleus*, xylem was initiated discontinuously at every node along the length of the stem and never in the internode (Bruck and Paolillo 1984 e). Also in *Coleus*, the strands were larger in transverse-sectional area and contained a greater number of vessels within the nodes than along the internodes. Plotkin and Aloni (unpublished) found that the vessel elements in the nodes of *Coleus* were shorter and narrower than in the internodes.

An experimental study of the hydraulic architecture of trees led Zimmermann (1978, 1982) to develop the constriction hypothesis. Xylem in the junction of a branch and stem, as well as at the base of a leaf (node), becomes a water transport bottleneck due to the narrow vascular elements in the junction. The narrow tracheary elements in the junctions give priority in the water supply to the leader and the main stem over the lateral stem appendages The nodes and junctions have a marked hydraulic constriction leading to hydraulic segmentation of the plant body (Zimmermann 1983).

An extreme example of segmentation was reported by Zimmermann and Sperry (1983). In the palm *Rhapis excelsa* narrow tracheids differentiate in the nodes and connect the wide metaxylem vessels of the leaves with the wide metaxylem vessels of the stem. The latter metaxylem vessels must remain functional for many years because the palm lacks a vascular cambium. Accordingly, the tracheids of the nodes protect the large metaxylem vessels of the stem from cavitation following leaf abscission.

A related observation is that the frequency of vessel endings is significantly higher in the nodes of young stems of *Vitis, Populus, Olea* (Salleo et al. 1984), and *Medicago* (Wiebe et al. 1984). The nodes are therefore considered as "safety zones", because gaseous emboli and fungal spores fail to pass through the endings.

A hydraulic safety zone was also reported at the base of barley roots (Luxová 1986), where numerous very small tracheary elements in the scutellar node connected the vessels of the stem with the vessels of the primary seminal root. Short tracheids and vessel endings were also observed at the base of the root. In *Zea* the wide vessels of the nodal roots were connected to the wide vessels of the stem through relatively narrow regenerative tracheary elements that had redifferentiated from parenchyma cells (Aloni and Plotkin 1985).

As mentioned previously, Aloni and Zimmermann (1983) suggested that the rate of xylem differentiation is controlled by the amount of auxin. High concentrations of auxin result in a rapid rate of differentiation, whereas low levels cause slow differentiation. Therefore, the duration of maturation of individual elements increases from the young leaves towards the root tip (point 4). Slow differentiation permits more cell expansion before secondary wall deposition, and therefore results in wide vascular elements (point 5). An interesting example of these points was recently studied in the nodal roots of *Zea*, which have an inner ring of large late-differentiated metaxylem vessels. The widest metaxylem vessels of a mature field-grown *Zea* plant are formed in the later-formed nodal roots that develop just prior to flowering (Hoppe et al. 1986). It is not surprising to find the widest vessels in the later-formed nodal roots because, at this stage of development, the young leaves, which are the main source of auxin, are at the greatest distances from the basal nodes from which the nodal roots develop. In addition, it was also found that the maturation of these widest metaxylem elements of the root is very slow and consequently results in immature vessel elements with living protoplasts at the lower part of the root up to 20–30 cm from the tip (St. Aubin et al. 1986). In younger and smaller *Zea* plants the "living" vessel elements were up to 10 cm from the tip (Higinbotham et al. 1972). McCully et al. (1987) have shown that these slowly differentiating vessel elements at the lower part of the wide metaxylem vessels in the nodal roots of *Zea* accumulate potassium from the soil. The "living" vessel elements near the root tip contain high concentrations of K^+ before they lose their cross walls. The mean K^+ concentration was about 200 mM, and the highest levels approached 415 mM. During vessel maturation the K^+ was released into mature vessels, and the ion moved upwards with the transpiration stream. The concentration of K^+ in mature vessels was about 15 mM.

3.8 Role of Gibberellin and the Control of Fiber Differentiation

Fibers are long narrow cells possessing thick secondary walls that are usually lignified, and their main function is support. Fibers are found in both phloem and xylem. Fibers arise from short initials that greatly elongate during their maturation. A novel type of differentiation, in which parenchyma cells between the longitudinal vascular strands redifferentiated to form regenerative phloem fibers (Fig. 3.7), was induced around wounds in young internodes of *Coleus* (Aloni 1976b). Regenerative fibers are relatively slow to differentiate and can be observed 3 weeks or more after wounding. They elongate mainly by intrusive growth at their apical and basal ends.

Spraying the leaves of intact plants with GA_3 increased the number of fibers in jute (Sircar and Chakraverty 1960), increased the length of fibers in hemp (Atal 1961), and increased the number and size of fibers in other genera (Stant 1961, 1963).

Excision-substitution experiments were needed in order to trace the stimuli that induce fiber differentiation. Sachs (1972b) studied primary phloem fiber dif-

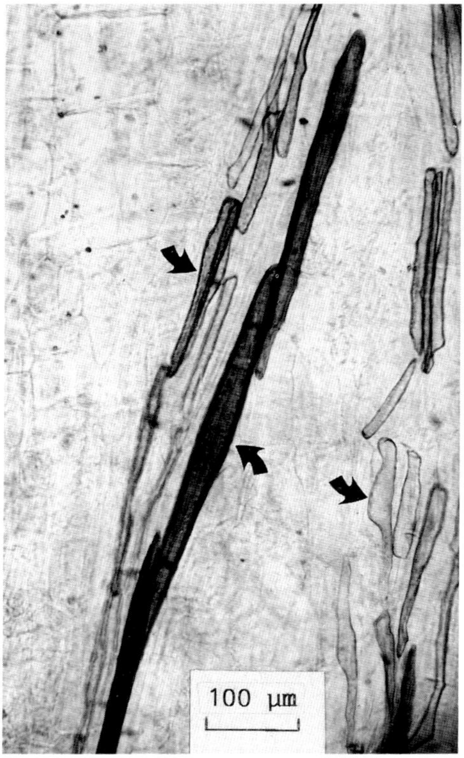

Fig. 3.7. Thick preparation, cleared with lactic acid and stained with lacmoid, showing regenerative phloem fibers induced by wounding a young internode of *Coleus blumei*. The regenerative fibers (*arrows*) have differentiated from parenchyma cells during a 60-day period following wounding. The largest fibers in the middle of the photomicrograph were strongly elongated. These fibers have increased in length about 20-fold compared to the size of the neighboring parenchyma cells. × 125. (Aloni 1976b)

ferentiation in the special strands of pea (Aloni and Gad 1982) and found that the differentiation of these strands depends on stimuli originating in very young leaf primordia. Early removal of these primordia prevented the formation of the fiber strands.

Studies of primary phloem fiber differentiation in *Coleus* have shown that the differentiation of fibers is induced by stimuli originating in young (Aloni 1978) as well as mature leaves (Aloni 1976c, 1978). Excision experiments, combined with wounding of stems of *Coleus*, have shown that the signals of fiber differentiation originate in the leaves and flow in a strictly polar fashion from the leaves to the roots (Fig. 3.8). Wounding experiments with pea revealed that the inductive stimuli travel polarly along the plant axis, and induce fibers along their pathway in both the stem and the root (Gad and Aloni 1984).

Aloni (1979) has shown that the role of the leaves in the differentiation of primary phloem fibers in *Coleus* can be replaced by exogenous application of combined IAA and GA_3 (Fig. 3.9). Both of these regulators have been found to be limiting and controlling factors in fiber differentiation. IAA applied alone caused the differentiation of a few fibers, whereas GA_3 alone did not exert any visible effect on fiber formation. Evidently GA_3 can affect phloem fiber differentiation only in the presence of auxin. Combinations of IAA with GA_3 completely re-

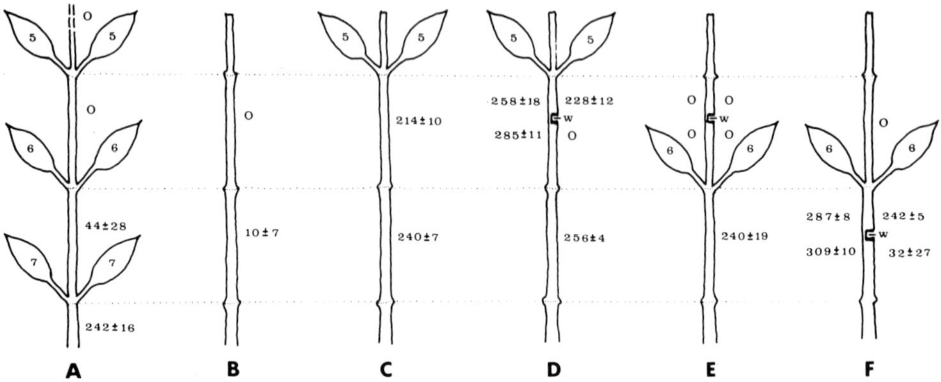

Fig. 3.8. Diagram illustrating the effects of various treatments on the differentiation of primary phloem fibers in stem internodes 5 and 6 of *Coleus blumei*. Numbers indicate the mean number of fibers (± standard error) in the adjacent half-transverse section in the middle of the internode, or 3 mm above and below a wound (−*W*). Plants of diagram *A* were examined at 0 time. All of the other treatments (*B–F*) had their lateral shoots and leaves excised at 0 time and were harvested after 14 days. (Aloni 1976c)

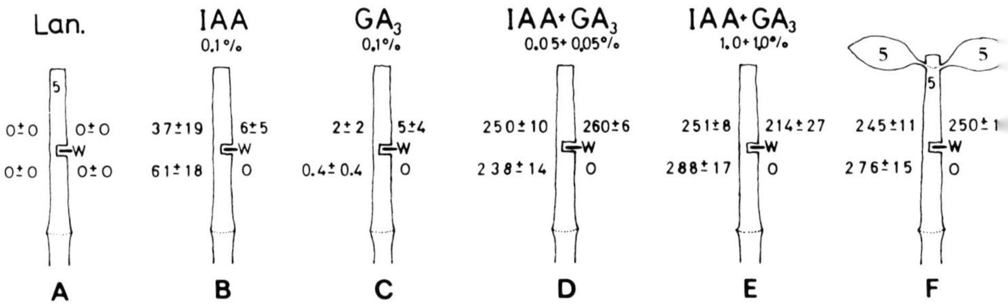

Fig. 3.9. Diagram illustrating the effects of IAA, GA_3, and the combination of both growth regulators in replacing the effect of the superjacent leaves in controlling the differentiation of primary phloem fibers around a wound (−*W*) in internode 5 of *Coleus blumei* stems. The growth regulators were applied in lanolin and all the plants (treatments *A–F*) were harvested after 10 days. (Aloni 1979)

placed the role of the leaves in phloem fiber differentiation both qualitatively and quantitatively (Aloni 1979).

The effect of the growth regulators on fiber differentiation, however, diminished considerably with increasing distance from the source of application (Aloni 1979). These findings probably explain the decrease in the number of fibers and the increase in their size along the plant axis from leaves to root apices (Denne and Whitbread 1978, Aloni and Gad 1982, Dodd 1985).

When various combinations of both growth regulators were applied in lanolin to decapitated and excised stems, high concentrations of IAA stimulated the rapid differentiation of fibers with thick secondary walls, whereas high levels of GA_3 produced long fibers with thin walls (Aloni 1979). The induction of long fibers by GA_3 is compatible with the findings of earlier studies on the effect of

GA_3 applied in the form of a spray to leaves of intact plants (Atal 1961, Stant 1963).

The combined presence of both growth regulators is also a requirement for the differentiation of secondary xylem fibers in *Populus* (Digby and Wareing 1966) and in *Phaseolus* (Hess and Sachs 1972).

The experiments with IAA and GA_3 lead to the conclusion that the differentiation of fibers along with the conducting elements in the vascular strands results from their common dependence on the polar movement of IAA. The difference, however, between the differentiation of the conducting elements and that of fibers is that the latter necessitates an additional stimulus, namely GA_3 (Aloni 1979).

Aloni (1982) has shown that cytokinin is required during the early stages of fiber differentiation, although the later stages of fiber maturation can occur in the absence of cytokinin. The correlation between the development of the plant body and the differentiation of its supportive tissues may arise from a common dependence on the same shoot/root feedback control signals (Aloni 1982). Indeed, experiments with intact *Helianthus* plants support this view and indicate that fiber differentiation in the secondary xylem is dependent on the inductive signal originating in the root apices, namely, on cytokinin (Saks et al. 1984).

3.9 Transfer Cells

Special vascular elements are found in sites where short-distance transport of high rates is required, as well as in the constriction of bottleneck regions where narrow vascular elements differentiatiate between wider vascular elements. Some of the parenchyma cells in the tissue surrounding the vascular channels develop secondary wall ingrowths (Fig. 3.10). These wall invaginations greatly increase the area of the plasma membrane, an increase which may be as high as 20-fold and probably results in a substantial increase in the transport capacity of the cells (Pate and Gunning 1972). These modified parenchyma cells are termed transfer cells (Gunning and Pate 1969, Pate and Gunning 1972) and are found in both phloem and xylem. However, transfer cells can be found in different plant tissues where intensive symplast-apoplast transport occurs (Gunning 1977). The wall ingrowths may develop over the whole periphery of the transfer cell or they may be restricted to the portions of the wall that associate with tracheary or sieve elements (Pate and Gunning 1969, Gunning and Pate 1969). Transfer cells usually differentiate in the minor veins of the leaves, developing from companion cells, and probably function in loading photosynthetic products into the sieve tubes (L. R. Peterson and Yeung 1975, Gilliland et al. 1984, Harris and Chaffey 1985). In addition, transfer cells are found at nodal regions of the shoot (Pate and Gunning 1972, Habricot and Sossountzov 1984) which are known to be restriction sites for transport and also considered hydraulic safety zones (Zimmermann 1983, Salleo et al. 1984). At these locations the transfer cells probably increase the lateral transport capacity of the node and compensate for the decrease in the flow rate of water through the narrow vascular elements and vessel endings of the nodes. In the vascular tissues of the roots, the transfer cells are restricted to the regions of lateral root junctions and to root nodules (Letvenuk and R. L. Peterson 1976, New-

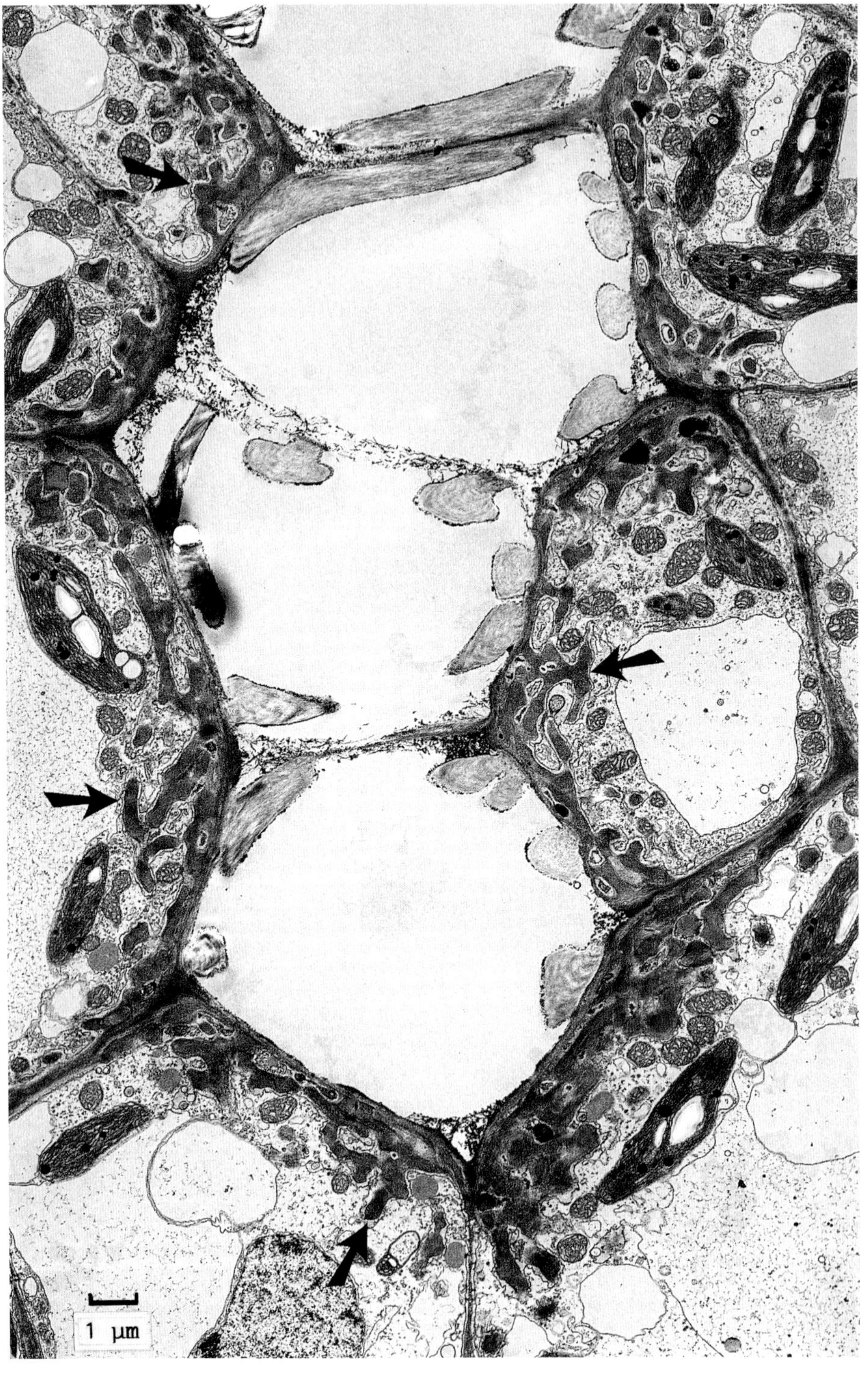
1 µm

comb and R. L. Peterson 1979). The formation of vascular transfer cells can be induced in the roots by wounding (Hardham and McCully 1982a) or following injury associated with nematode infection (M. G. K. Jones and Northcote 1972, M. G. K. Jones et al. 1974). Induction of wall ingrowth formation in phloem transfer cells in pea leaves is promoted by light or by the presence of soluble carbohydrate (Henry and Steer 1980). Harris and Chaffey (1985) described the distribution of plasmatubules, which are tubular invaginations of plasmalemma, in transfer cells of the leaf veins of pea associated with both sieve and tracheary elements. They suggested that plasmatubules in transfer cells may act as an additional structural modification for symplastic uptake into transfer cells at sites where the apoplastic supply of solutes is at its maximum. There are no hypotheses concerning the control of vascular transfer cell formation. It is reasonable to suppose that growth regulators influence their differentiation, but there are no studies relevant to this possibility.

3.10 Vascular Adaptation

The influence of environmental factors on wood structure of living and fossil trees was reviewed by Creber and Chaloner (1984). Additional effects of various physical factors on phloem and xylem differentiation are discussed in Chapter 5. However, there is not much knowledge on mechanisms of adaptation of vascular systems of plants to their environments and the possible role of plant growth regulators in vascular adaptation.

Carlquist (1977) proposed the concepts of "vulnerability" (V) and "mesomorphy" (M) indices. Vulnerability indicates the possibility of cavitation occurring in vessels. The V index is calculated by dividing the mean vessel diameter by the number of vessels per mm^2. A low V value indicates a great "redundancy" of vessels. The more numerous the vessels, the lower is the probability that water conduction would be seriously damaged by air embolism formed under water stress. The mesomorphy index is derived by multiplying the V index by mean vessel element length. Conditions that favor better stem growth are expected to result in longer vessel elements. In a study conducted in southwestern Australia, Carlquist (1977) found positive correlations between the V and M values and the environmental water supply of the plants studied. Mesophytes were found to have low V and high M values, whereas xerophytes had high V and low M values. Similar general trends were reported for Oleaceae species from China, where the taxa from mesic (subtropical) provenances tend to have longer vessel elements and wider and less frequent vessels than their relatives from more temporate or seasonally dry habitats (Baas and Xinying 1986).

Fig. 3.10. Epithelium of xylem parenchyma transfer cells surrounding a file of tracheary elements in a leaf trace strand within a node of *Anagallis arvensis*. The wall ingrowths (*arrows*) are restricted to the portions of the walls that associate with tracheary elements. ×7000. (Gunning 1977, Fig. 11)

Fig. 3.11. Effect of pot size on the diameter and secondary wall thickness of the vascular elements formed in the stems of *Hibiscus cannabinus* plants. The experiment was terminated after 60 days. (**A**) Development of the plants after 60 days in large (*b*), medium (*c*), and small (*d*) pots. The plant in the large pot grew under the most favorable conditions and was the largest. Those in the smaller pots were subjected to different levels of stress that limited the size of the plants. Photomicrographs (**B, C, D**) show portions of transverse sections of the same magnification and in the same orientation. The sections were taken from the middle of the first internode above the co-

Comparative ecological wood anatomy studies done in Israel and California have shown that plants grown in extreme habitats usually have high densities of very narrow vessels, which are considered as safe systems adapted to drought conditions (Baas et al. 1983, Baas and Carlquist 1985, Carlquist and Hoekman 1985, Fahn et al. 1986). On the other hand, plants grown in mesomorphic environments have low densities of wide vessels which are more effective in water conduction.

Zimmermann (1983) emphasized that vessels (consisting of a series of vessel elements forming together long capillaries limited in length by imperforate walls at both extremities) and not vessel elements, are the operating conduction units of the xylem in angiosperms. Therefore, vessel length should be considered when the safety of the water system is studied. The safety of water conduction is markedly decreased as a result of increase in vessel diameter and vessel length. On the other hand, the efficiency of water conduction is substantially increased with increasing vessel diameter and length. In ideal capillaries conductivity is proportional to the fourth power of the diameter (Zimmermann 1983). Therefore, at a given pressure gradient, the relative volumes of water flowing through capillaries of diameters 1, 2, and 4 are 1, 16, and 256, respectively. In order to emphasize how safety is related to efficiency of water conduction, a narrow-vessel tree, such as maple (*Acer saccharum*) is compared with a wide-vessel tree, such as oak (*Quercus ruba*). The first is a diffuse-porous tree with vessel diameter of about 75 µm and vessel length of 30 cm. The oak is a ring-porous tree with earlywood vessels that have diameters of about 300 µm and a length of 10 m. Taking vessel width and length into consideration, the damage done (air embolism) to one earlywood vessel in the oak is 256 × 30 (7680) times more serious than in the maple. A trunk of a maple needs, however, about 7680 times as many vessels as does the similarly sized trunk of an oak (Zimmermann 1983). These numbers illustrate the contrasting needs of safety and efficiency.

Aloni (1987b) suggested that the environment controls the dimensions of the vessels and their distribution pattern through its control of the development and height of the plant. The height and rate of branching determines levels of the polar flow of auxin along the plant and probably other growth regulators. In order to explain how ecological conditions control the size and frequency of vessels and fibers in plants, Aloni (1987b) proposed the following vascular adaptation hypothesis: (1) Limiting conditions in the environment limit the final size of the plant and result in small and suppressed shoots, whereas conditions which do not limit plant development allow the plant to reach its maximum potential height and size. (2) The duration of the growth period determines total plant development. In extreme and limiting habitats the active growth period is relatively short and results in small plants, while stable and moderately comfortable conditions

tyledons. The photomicrographs (**B, C, D**) correspond to the different treatments (*b, c, d*). The results of the experiment indicate that a decrease in pot size, and consequently of plant size, resulted in the following anatomical changes: (*1*) A decrease in the diameter of the secondary vessels (*v*) and secondary xylem fibers (*xf*); (*2*) an increase in the wall thickness of secondary xylem fibers; (*3*) an increase in vessel density (number of vessels per unit of transverse-sectional area of xylem), vascular cambium (*c*), and phloem fibers (*pf*). (R. Aloni, P. Feigenbaum and T. Plotkin, unpublished) allow more growth activity throughout the year. Thus, more growth is accomplished and the result are large and well-developed plants. (3) The height of the plant and the rate of its branching determines gradients of auxin along the axis of the plant. Increases in plant height and decreases in branching increase the gradients of auxin from the young leaves (sources of auxin) to the lower parts of the stem. In small shrubs, which are typical of both cold and dry habitats, as well as locations where there is not enough soil for the roots, the distances from the young leaves to the roots are very short and no substantial gradient of auxin can be formed. Therefore, the levels of auxin are relatively high along these small plants and result in the differentiation of numerous very small vessels in the greatest densities, as expected by the six-point hypothesis (Aloni and Zimmermann 1983). These plants also have small fibers with thick secondary walls (Aloni 1979). On the other hand, in large trees and in long lianas, the very long distances from the young leaves to the roots enables a substantial decrease in auxin levels in the lower parts of the stem and in the roots, which result in the differentiation of very wide vessels in lower densities. These locations are also characterized by large fibers with relatively thin secondary walls.

Although there is some experimental evidence that supports the above hypothesis (Aloni 1987b) (Fig. 3.11), it should be tested in various experimental systems and under different growth conditions. A recent comparative study of the xylem in the genus *Frankenia* (Whalen 1987) supports the adaptation hypothesis (Aloni 1987b) by showing that there is a general tendency for the dimensions of the axial wood elements to be positively associated with the plant height. In general, differences in wood anatomy more strongly reflect differences in plant growth form and size than phylogeny (Whalen 1987).

Chapter 4

Evidence from Wound Responses and Tissue Cultures

L. W. ROBERTS

4.1 Rationale for Using Wounds and Tissue Cultures to Study Vascular Differentiation

Vascular wounds of intact plants and experiments with isolated and cultured tissues have certain advantages in studies involving cytodifferentiation. By using the entire plant or plant organ, polarity inherent in the shoot-root axis is maintained and the spatial arrangement of the regenerated phloem and xylem can be examined in relation to the existing vascular system. Sachs (1984) has used this technique with excellent results. Because of the correlative effects arising from the various interconnected organs, many workers have preferred to study cytodifferentiation in isolated fragments of tissue and in single cells. These cellular systems, devoid of vascular tissues associated with the parent plant, mainly consist of parenchyma. Although parenchyma may give the appearance of consisting of relatively homogenous cells, structural, physiological, and biochemical gradients were demonstrated in *Nicotiana* pith parenchyma (Lavee and Galston 1968). These gradients may have different potentials for the induction of cytodifferentiation, that is, to auxin sensitivity (Trewavas 1982). The sequence of inductive events can be timed as they occur in a primary explant taken directly from the parent plant and in wound responses, that is, by the reprogramming of quiescent cells. For example, cell cycle activity prior to any visible signs of cytodifferentiation has been examined (Dodds 1981 b), as well as observations on changes in microtubule orientation during tracheary element differentiation (Falconer and Seagull 1985 a, b, 1986). Both organ wounding and tissue excision induce ethylene biosynthesis which may contribute to cytodifferentiation (Chap. 5.5, 5.6). Organ wounds are based on severing vascular continuity within the organ. This operation disturbs the normal movement of growth regulators and causes the release of endogenous hormones. Leakage of hormones around the wound initiates the redifferentiation of parenchyma into phloem and xylem elements, and, in addition, starts cell division leading to the establishment of a wound cambium. Phloem elements are the first to appear and may be formed in the absence of regenerated xylem. Some workers have attempted to remove endogenous hormone sources by disbudding, defoliation, and decapitation, whereas others have used the vascular system as a conduit for feeding hormones into the wound site. With isolated and cultured plant tissues the attempt is made to minimize endogenous hormones; auxin and cytokinin are incorporated into the nutrient medium. A problem exists with the presence of endogenous growth regulators within the primary explant as well as biosynthesis of hormones during the experimental period. Developmental changes in the content of growth factors are probably involved in the observation that mature and immature tubers of Jerusalem artichoke show different types of xylogenic responses (Phillips 1981 b). Cell suspensions, and the

Zinnia system in particular, provide excellent material for biochemical studies on cytodifferentiation. In the present discussion nutritional aspects of vascular tissue differentiation in culture will be omitted, unless nutrition has some bearing on hormonal physiology.

4.2 *Coleus* Stem Wounds and Cultured Internodal Explants

All students of developmental botany are familiar with the studies on *Coleus* stem wounds conducted by W. P. Jacobs in the 1950s (reviewed by W. P. Jacobs 1984). Auxin was established as the limiting hormone in wound xylem differentiation in this system, and W. P. Jacobs (1959) devised the PESIGS criteria for determining the critical chemical variables in a cytodifferentiation event. W. P. Jacobs' experiments also suggested the idea of studying xylogenesis by using internodal slices of *Coleus* cultured on a xylogenic medium (Fosket and Roberts 1964, Fosket 1968, 1970, 1972).

The question often arises concerning the functional nature of vascular tissues regenerated around wounds. Two early workers, using different techniques, studied this question and came to the conclusion that wound regenerated vascular tissue is functional in conduction (von Kaan Albest 1934, Eschrich 1953). Recently Schulz (1987) examined conduction in sieve elements regenerated in *Pisum* roots after complete severance of the stele. As indicated by positive staining of the sieve plates with aniline blue, a continuous chain of sieve elements had differentiated around the wound 55–62 h after wounding. Effective translocation, as visualized by fluoresceine as a phloem-mobile marker, did not occur until approximately 10 h later. One should not consider that all cells observed in tissue cultures that bear a resemblance to tracheary elements are necessarily functional in water conduction. In addition to cells that apparently are tracheids and vessel elements, cultures generally contain a host of sclereid-like and fiber-like abnormalities that defy categorization.

Aloni has conducted numerous studies on vascular formation in *Coleus* stems (1976a, b, c, 1978, 1979, Aloni and W. P. Jacobs 1977a, b). One study followed the timing of sieve element and tracheary element differentiation in relation to the presence of phloem anastomoses (Aloni and W. P. Jacobs 1977b). In wounded mature internodes, a few sieve tubes and vessels formed after 3 days, and the number of regenerated vascular elements sharply increased by 4 days. The number of sieve tubes and vessels formed after 4 days was closely related to the number of preexisting bundles of the type of vascular tissue that had been severed by the wound. The ratio of bundles severed (phloem : xylem) was 2.14, and the ratio of regenerated sieve tubes to vessels was 2.24. In another study, Aloni (1976c) found that primary phloem fiber initials differentiate in the periphery of the vascular bundle adjacent to the primary sieve elements. This differentiation process is dependent on stimuli originating in the leaves, and the stimuli flow along the stem in a basipetal direction (Sachs 1972b, Aloni 1978). Since the polar movement of auxin in the *Coleus* shoot induces the differentiation of sieve tubes or a combination of sieve tubes and vessels (LaMotte and W. P. Jacobs 1963, Thompson and W. P. Jacobs 1966, W. P. Jacobs 1970), an additional polar stimulus flow-

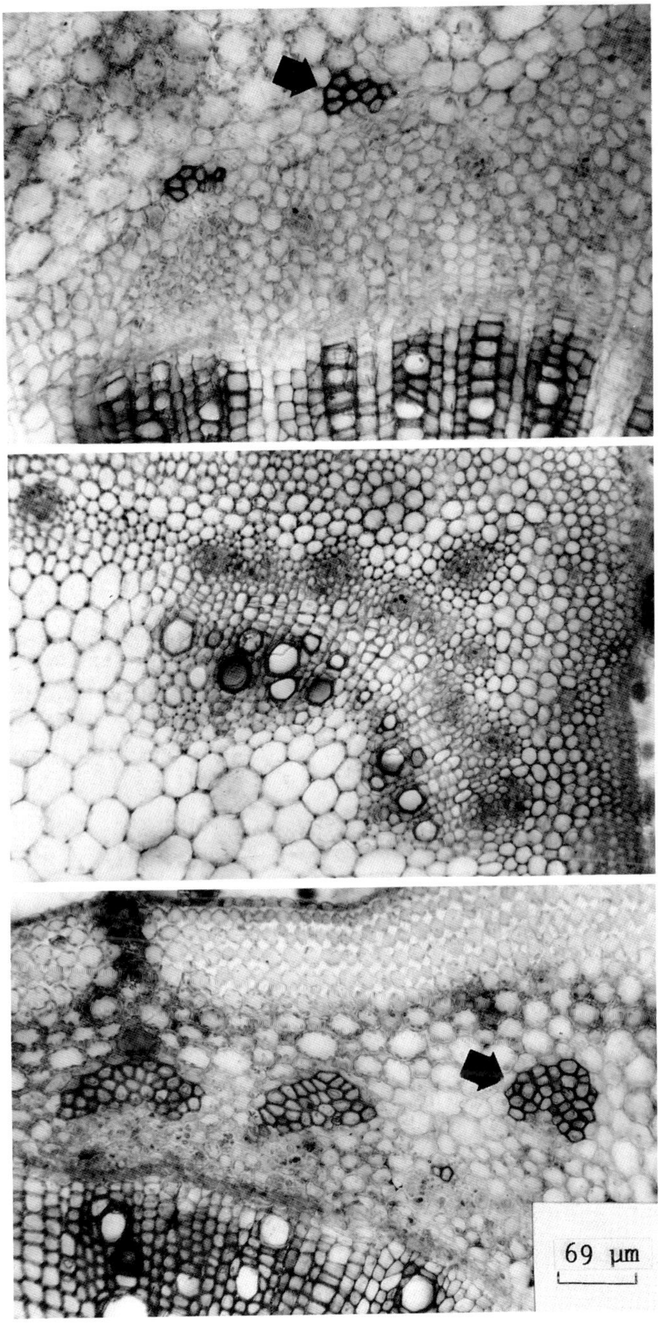

Fig. 4.1. Transverse sections taken from the middle of the first internode of *Coleus* plants showing the pattern of primary phloem differentiation induced by various hormone treatments. IAA and GA_3 were applied apically in lanolin paste to disbudded and defoliated shoots. *Arrow* indicates the presence of primary phloem fibers; bar = 69 µm. *Upper* Small bundles of primary phloem fibers induced by IAA (0.1% w/w). *Center* No fiber differentiation occurred following the application of GA_3 (0.1% w/v). *Lower* Heavy production of primary phloem fibers resulted from the combined treatment of IAA (0.05% w/w) plus GA_3 (1.0% w/v). (Aloni 1979)

ing through the cells in the periphery of the bundle should regulate the formation of primary phloem fibers. Aloni (1979) assumed that this additional stimulus was gibberellic acid. The hypothesis was tested, and the results indicated that the combined effect of IAA and GA_3 completely replaced the role of the leaves in the differentiation of primary phloem fibers (Fig. 4.1). Aloni (1979) found that the final size of the primary phloem fibers was determined by the IAA:GA_3 ratio. High levels of IAA induced short fibers with thick secondary walls, whereas high levels of GA_3 yielded long fibers with thin walls. Size of the phloem fibers was correlated with the overall dimensions of the internode, indicating that both hormones exert an influence on the development of the stem internode. As mentioned previously (Chap. 2.4), the technique of increasing the commercial production of fibers by treating fiber crops with a combination of NAA and GA_3 has been patented by Aloni.

Excision of a leaf of *Coleus* before or early in the differentiation of leaf traces severely reduced vascular differentiation in the subjacent internode (Bruck and Paolillo 1984e). The application of anion exchange beads loaded with IAA completely restored the formation of leaf traces with regard to arrangement, composition, size, and rate of development. Plants treated in this manner with exogenous IAA exhibited xylem differentiation in association with putative phloem-only bundles. Phloem-only bundles probably represent an auxin-deficient stage in the early development of collateral bundles, and the developmental process is dependent on auxin availability to the bundle for the differentiation of both types of vascular tissue (Bruck and Paolillo 1984e). These authors have conducted detailed studies on the vascular anatomy of the *Coleus* shoot system (Bruck and Paolillo 1984a, b, c, d).

A study was made of the ultrastructure, division pattern, and course of wound phloem differentiation in *Coleus* by Behnke and Schulz (1980). Wound phloem bridges developed 6 days after wounding. These regenerated structures contained sieve elements, companion cells, and phloem parenchyma cells. Three consecutive divisions occurred prior to the formation of phloem mother cells. Companion cells were obligatory, sister cells to wound sieve elements, and connected to the latter by plasmodesmatal strands. Most of the regenerated sieve elements were still nucleated 6 days after wounding, although they already contained P-protein bodies and plastids filled with sieve element starch. Their cytoplasmic differentiation corresponded closely to the changes recorded during the maturation of normal sieve elements. Wound sieve elements were not connected to preexisting bundle sieve elements, that is, they opened a new tier of young sieve elements produced by cambial activity, as reported earlier by Eschrich (1953) and Benayoun et al. (1975). There are two main differences between the ontogeny of wound sieve elements and bundle sieve elements. Plastids in wound-regenerated cells redifferentiate from preexisting chloroplasts or amyloplasts, whereas in ordinary sieve elements plastids develop from meristematic proplastids. Another difference is that sieve pores of the wound cells differentiate from preexisting plasmodesmata located in primary pit fields connected to the original parenchyma cells (Behnke and Schulz 1980).

Explants of *Coleus* pith parenchyma were used to examine xylogenesis by Earle (1968) and Comer (1978). Both workers agreed that the only exogenous requirements for the induction of xylogenesis were an auxin and a carbohydrate. Earle (1968) found that the addition of kinetin was either inhibitory or without any effect when the auxin-induced response was optimal. In two experiments, however, the addition of kinetin increased the xylogenic response in the presence of a suboptimal level of IAA (10^{-5} M). Since it is unlikely that these parenchyma cells initiated xylogenesis without a source of cytokinin, it can be assumed that either the cells synthesized the hormone, or endogenous cytokinin came from vascular tissue during the isolation procedure. Neither worker mentions rinsing the pith samples after removing the vascular tissues. Comer and Leonardo (1981) found that explants (unrinsed) of pith isolated from *Coleus* differentiated sieve elements in response to a combination of IAA and sucrose. Although increased sucrose concentrations enhanced phloem formation, the level of auxin was not a factor in stimulating the process. The optimal levels were 1.0 mg l^{-1} combined with 4% (w/v) sucrose. Kinetin, benzylaminopurine, and GA_3 were ineffective in the induction of phloem differentiation. Different results were obtained by Houck and LaMotte (1977). These workers found that both auxin and zeatin riboside were necessary for the induction of sieve element differentiation. Phloem regeneration occurred in the absence of wound xylem formation, as a result of severing of phloem-only bundles in the fifth internode of *Coleus*. Phloem regeneration is evidently not dependent on some factor released from severed xylem. Feeding auxin to a phloem-only bundle that has been severed will induce only sieve elements and not tracheary elements. Phloem regeneration was examined in plants that had been disbudded, defoliated, and decapitated. The application of IAA in lanolin completely restored phloem development to the level found in the intact plant. IAA, however, failed to induce phloem formation in the absence of the root system. Phloem regeneration was restored in isolated internodes by using an apical application of IAA plus a basal treatment of either zeatin or zeatin riboside. Since phloem regeneration required the combined effects of auxin and zeatin, apparently the phloem of the intact plant is carrying endogenous zeatin from the root system. As tracheary elements are also induced to form in the presence of auxin and cytokinin, it is puzzling that these vascular elements are not differentiated at the same time.

4.3 Root and Shoot Wounds of *Pisum*

Several investigators have examined cytodifferentiation occurring in wounds that rupture the stelar conducting tissues of primary roots of *Pisum sativum* (Robbertse and McCully 1979, Hammersley and McCully 1980, Hardham and McCully 1982a, b, Rana and Gahan 1983a, Schulz 1986a, b, 1987). The quiescent cortical cells of pea roots can be induced, by severing the vascular cylinder of the root, to initiate cytodifferentiation to form files of tracheary elements and sieve elements (Fig. 4.2). Vascular transfer cells are differentiated at the same time. These developmental processes are accompanied by cell division. The planes of cell division, particularly in relation to subsequent sieve element formation, are precisely aligned in contiguous cells. Thus the regenerated conducting cells are arranged in smooth files (Robbertse and McCully 1979).

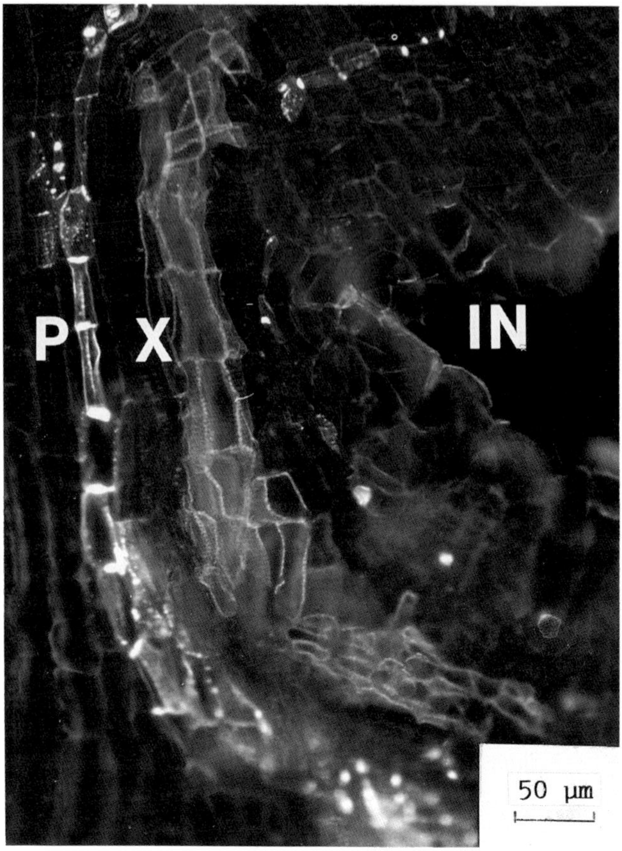

Fig. 4.2. Regeneration of phloem (*P*) and xylem (*X*) bridges around the severed stele of a primary root of *Pisum sativum* 3 days following wound incision (*IN*). The base of the root is toward the top of the photomicrograph. The section was mounted in 0.05% (w/v) aniline blue and photographed through fluorescence optics; bar = 50 µm. (Robbertse and McCully 1979)

Hammersley and McCully (1980) examined wound xylem differentiation in pea roots in the presence of colchicine. Since colchicine binds to tubulin protein, the treated cells lacked a spindle, a phragmoplast, and microtubules. Xylogenesis was delayed in the presence of colchicine (78 h post-wounding compared to 48 h in untreated controls), and the authors concluded that xylogenesis occurred directly from cortical parenchyma without an intervening cytokinesis. Similar results were reported by Hardham and McCully (1982b). Vascular differentiation occurred in the presence of caffeine as well as colchicine. Caffeine (1,3,7-trimethyl xanthine) blocks cell plate formation following karyokinesis by preventing the fusion of vesicles in the mid-zone of the phragmoplast (Becerra and López-Sáez 1978, M. G. K. Jones and Payne 1978). Bi- and multi-nucleate cells are formed in the presence of both drugs (Hardham and McCully 1982b).

Gad and Aloni (1984) studied primary phloem fiber differentiation in root and stem of pea. The deposition of lignified cell walls was first observed after 2 and 5 days, respectively. Wounds in both root and stem interrupted the differentiation of fibers directly below the wounds. Although the hormonal requirements for fiber differentiation in pea roots are unknown, the early stages of fiber development in cultured *Helianthus* hypocotyl segments (Aloni 1982) and in young

Helianthus plants (Saks et al. 1984) appear to be regulated by cytokinin synthesized in the root apices. The wounding experiments in pea have indicated that fiber differentiation in both root and shoot are regulated by polar inductive stimuli, probably IAA and GA_3 (Aloni 1979), arising from the direction of the leaves.

The wound response in pea roots was employed to study cell determination by Rana and Gahan (1983a). Quantitative cytochemical observations of the cortical parenchyma cells following rupture of the stelar vascular system revealed the onset of esterase activity. This enzymatic activity was an early indication of the determination of these cells to form a vascular bridge. An esterase bridge was first observed 18 h after wounding, and xylem elements started to differentiate in the same position about 72 h after wounding. Explants were removed at various times from the wounded roots and incubated in a basal medium containing sucrose but lacking exogenous hormones. A minimum of 8–10 h was required before explant removal from the parent plant in order to produce detectable esterase activity. These same tissues must remain associated with the pea seedling for a minimum of 19–20 h before the cells are capable of differentiating xylem elements. Thus, although genetic information is available as early as 8–10 h after wounding, the information does not appear to be expressed until several hours later. A further study of the relationship of esterase activity to xylogenesis was undertaken by Gahan et al. (1983). Explants incubated on a basal medium for 3–6 days displayed esterase activity similar to the staining reactions given in sections of cortex and stele from intact roots of a similar age. In the presence of either auxin or cytokinin there was a three- or twofold increase in enzyme activity in the cortical parenchyma, respectively. In the presence of both hormones, that is, a xylogenic combination, there was a threefold increase in enzyme activity. In all experiments the enzyme activity was completely blocked in the presence of either diisopropylfluorophosphate or diethyl *p*-nitrophenylphosphate. Inhibition of the staining reaction by these reagents is an indication of carboxylesterase (EC 3.1.1.1) activity. The low level of activity normally found in cortical cells was resistant to the inhibitors, indicating acetylesterase (EC 3.1.1.6) activity. Auxins and cytokinins appear to activate similar carboxylesterases during xylem differentiation from cortical parenchyma cells of pea (Gahan et al. 1983).

The developmental anatomy of wound phloem regeneration in primary roots of pea following the rupture of the stele has been described by Schulz (1986a, b). Schulz (1987) also examined the transport capability of wound sieve elements (Sect. 4.1).

Two general points should be raised with respect to the involvement of hormones in the formation of a vascular bridge. A model can be proposed in which the timings of the appearance of the first genetic changes could be explained by a simple diffusion model involving auxin, cytokinin, and sucrose (Rana and Gahan 1983a). The first point concerns an examination of the early stages of bridge formation which indicates that it is initiated from both ends at the same time, so that if auxin and cytokinin are the prime movers, they must be available in the correct ratios at both ends of the bridge at the same time. Additional work is needed to show if this is the correct scenario and what are the mechanisms involved. The second point relates to the fact that programming for carboxylesterase activity and secondary cell wall biosynthesis occurs well in advance of mitosis in the bridge-forming regions. The implications are that a quantal mitosis is not a preliminary requirement for cytodifferentiation (Chap. 1.5), and that such cell division is primarily concerned with increasing the tissue mass that will form the ultimate vascular bridge. This also raises the interesting possibility that the priming of the parenchyma cells to form vascular elements occurs adjacent to cells which are being primed to divide. It is uncertain which hormones are involved in each of these events and in which manner they are organized to yield spatially separate events.

4.4 Vascular Differentiation in Other Wound Systems

Tracheary elements have been observed bridging wounds in the stem of *Zea mays*, a monocotyledon (Aloni and Plotkin 1985). Xylem formation around the wound was relatively slight, limited to the youngest internodes, and declined sharply in a basipetal direction. The regenerated tracheary elements, largely confined to the region above the wound site, were characterized by the presence of helical and an-

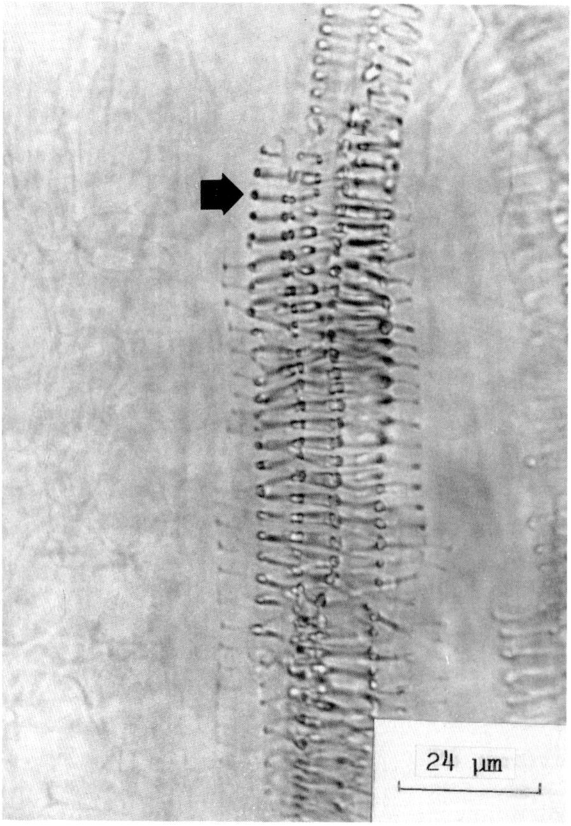

Fig. 4.3. Xylem regeneration in a young stem internode of *Zea* 7 days after wounding. Annular and helical secondary wall thickenings occur in the wound vessel elements. *Arrow* indicates annular thickening; bar = 25 µm. (Aloni and Plotkin 1985)

nular patterns of secondary wall formation (Fig. 4.3). Wounding produced vascular anastomoses in the leaf immediately above the wound site and discontinuous xylem elements in adjacent bundles. Aside from the wound response, regenerative tracheary elements were seen at the junction of adventitious roots with the stem internode, and these cells had pitted and reticulated secondary wall thickenings.

Several studies have been made on xylem regeneration resulting from stem wounds in gymnosperms. Wound effects in *Pinus taeda* were studied by Kuroda and Shimaji (1984 a) following the insertion of a pin (400 μm diameter) into the cambial region of the stem. Ray parenchyma cells proliferated at random in the wound zone, and some of these cells invaded mature tracheids through the pits. This resulted in the formation of tylosis-like structures. Subsequently the walls of the proliferated cells became reticulately thickened and lignified. Although the ray cell derivatives resembled tracheary elements observed in callus cultures, the investigators assumed that the proliferated cells had died prematurely for some unknown reason. In their opinion, these cells were not tracheary elements. Immature axial and ray tracheids in the vinicity of the wound remained determined and differentiated into tracheids, with some modifications in the structure of the bordered pits. Additional studies have been reported on the application of the pinning method as a marker for xylem growth in conifers (Yoshimura et al. 1981, Kuroda and Shimaji 1983) and hardwood species (Kuroda and Shimaji 1984 b). In addition, Kuroda and Shimaji (1985) studied wound effects following pin insertion in hardwood xylem of poplar (*Populus euramericana*). Parenchyma cells redifferentiated into tracheary elements around wounds in seedlings of *Pinus pinea* (Aloni and Saks 1984). The regenerated elements contained bordered pits and were shorter and wider than normal tracheids. Xylogenesis occurred basipetally and was first observed 3 days after wounding. The differentiation response required the presence of leaves above the wound site, and the application of NAA in lanolin replaced the leaves in the induction of xylogenesis. Evidently the hormonal stimulus for the induction of xylogenesis is the same in both dicots and gymnosperms (Aloni and Saks 1984). Savidge and Wareing (1981 a) have postulated the presence of a tracheid-differentiation factor in pine needles operating through an interaction with IAA (Ewers and Aloni 1985). Denne (1977) examined wounding in *Picea sitchensis* cambium resulting from a needle puncture. Tracheids in the wall-thickening stage were found to retain their cytoplasmic contents, and this may provide a means for determining seasonal changes in the duration of the wall-thickening stage. Cells in the wall-thickening stage further from the wound site had thicker walls, presumably due to the release of growth promoters from the wound site. Kucera (1977) found modified tracheids with an additional wall layer in the vicinity of the wound tissue in *Pseudowintera colorata*. This new layer appeared to be a repetition of the secondary S_2 and S_3 layers; it may be due to an increase in IAA released by the injured cells.

Rugenstein (1982) used a palladium microprobe (250 μm diameter) for wounding peduncles and petioles in *Gossypium hirsutum*. After a 2-week period there was a restoration of the vascular connections between the original wounded vascular tissues and the regenerated vascular system. Connections were not found between old and new vascular tissues in wounded pea plants by another group (Benayoun et al. 1975).

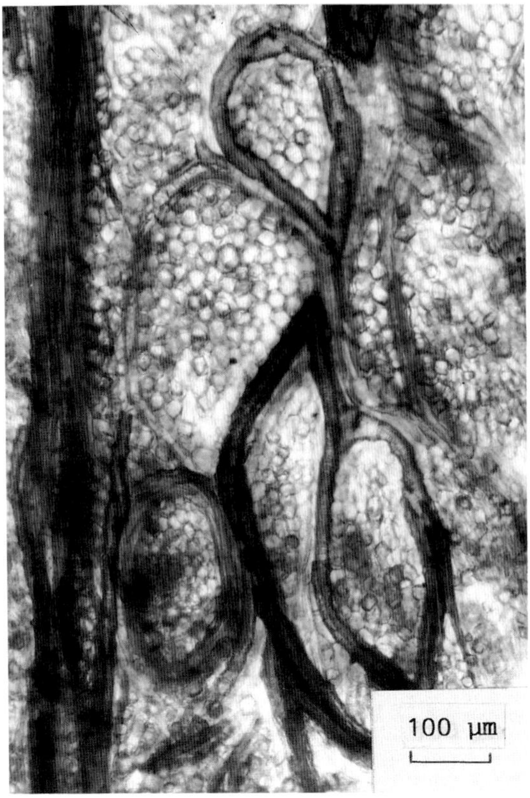

Fig. 4.4 Tangential section at the base of a suppressed bud located in the outer bark of *Ficus religiosa* tree contains an arrangement of closed circular vessels; bar = 100 μm. (Aloni and Wolf 1984)

The occurrence of vessels and sieve tubes in the form of rings serves to illustrate the control of vascular patterns by auxin flux. These unusual formations are found in certain types of wounds, grafts, untreated storage organs such as kholrabi (*Brassica caulorapa*) and Jerusalem artichoke (*Helianthus tuberosus*), and suppressed buds across the bole in *Ficus religiosa* (Aloni and Wolf 1984). Circular formations apparently occur in response to the circular transport of auxin and possibly other factors, that is, the pooling of these substances in a parenchymatous tissue. The formations are found in downward-pointing stem flaps on wounded bean seedlings (Sachs and Cohen 1982). Root initiation on stem cuttings reduced circular vessel formation, but the formation was increased when rooting was inhibited by the application of benzylaminopurine. Circular vessels were found in inverted grafts in which tissues with opposite polarity were contiguous, and at the base of embedded and suppressed buds (Aloni and Wolf 1984, Fig. 4.4). According to Sachs and Cohen (1982) cells are induced to differentiate because of a flux rather than a concentration gradient of the hormones.

4.5 Induced Xylogenesis in *Lactuca* Pith Parenchyma

Since Dalessandro and Roberts (1971) introduced the use of pith parenchyma explants excised from lettuce heads, numerous investigators have used this technique to study induced xylogenesis in vitro (Roberts 1976). A. R. Miller and his colleagues have provided evidence that ethylene has a synergistic effect on xylogenesis in this system (A. R. Miller and Roberts 1984, A. R. Miller et al. 1984, 1985, Chap. 5.5, 5.6). Lettuce pith contains various gibberellins and GA-like substances that may play a role in these in vitro experiments (Pearce et al. 1987, Chap. 2.4). There is evidence that auxin-induced xylogenesis in lettuce pith involves calmodulin as a second messenger (Roberts and Baba 1987, Chap. 2.2).

J. Warren Wilson et al. (1982) devised a sandwich technique in which cylinders of lettuce pith were fed various substances at opposite ends. Applications of IAA, zeatin, and sucrose in various combinations showed that callus was initiated with IAA, whereas xylogenesis required both hormones. Both callus and cytodifferentiation were restricted to a few millimeters at the ends of the explants. Presumably sucrose or its metabolic products diffused the entire length of the cylinders, and the hormones apparently remained at the site of application. Some callus and xylogenesis occurred in the absence of exogenous sucrose. There was some indirect evidence that high sucrose concentrations were inhibitory to xylogenesis (J. Warren Wilson et al. 1982).

A subsequent study was made on determination of tracheary element differentiation in lettuce pith cultures by the Australian scientists (Tucker et al. 1986). The medium was based on the MS salt mixture and vitamins, sucrose (1% w/v), agar (0.6% w/v), with glutamine and glutamate (10 mM each) replacing the nitrate and ammonium as nitrogen sources. The inductive medium contained IAA (10 mg l^{-1}) and zeatin (1 mg l^{-1}). After various intervals of culture on the inductive medium, the explants were transferred to a noninductive medium. Two types of noninductive media were used. low IAA (0.1 mg l^{-1} IAA, 1 mg l^{-1} zeatin) and low zeatin (10 mg l^{-1} IAA, 0.01 mg l^{-1} zeatin). The transfer experiments revealed that very few tracheary elements were formed in explants cultured 2 days on the inductive medium followed by 5 days on either of the noninductive media. Explants transferred to the low-zeatin medium on the third day produced about 6000 tracheary elements per explant. On the other hand, high numbers of tracheary elements were not formed in explants transferred to the low-auxin medium until the explants had remained for at least 5 days on the inductive medium. The results suggested that cytokinin-dependent determination events were completed earlier than processes involving auxin (Tucker et al. 1986).

4.6 Xylogenesis in Explants from Storage Organs

Numerous investigators have used explants and callus obtained from tubers of Jerusalem artichoke (*Helianthus tuberosus*) to study tracheary element differentiation (Gautheret 1961 a, b, Dalessandro 1973, Minocha and Halperin 1974, Phillips and Dodds 1977, Phillips 1981 a, b, Watson and Halperin 1981, Phillips and

Arnott 1983, Hawkins and Phillips 1983, Minocha 1984, Phillips and Hawkins 1985, Phillips 1987). Inconsistencies between the results from different groups may be partially due to differences in the levels of endogenous hormones in the parent tubers. Dormancy-regulating hormones are present within the tubers, and the tubers are known to contain an endogenous cytokinin (J. P. Nitsch and C. Nitsch 1960, Dalessandro 1973). Another difficulty in comparing results from different laboratories is that the cultural techniques were not the same. Some workers have suspended explants in liquid cultures (Minocha and Halperin 1974, Watson and Halperin 1981), others have used filter paper discs impregnated with a liquid medium (Phillips and Dodds 1977), and still others have used agar as a matrix (Gautheret 1961 a, b, Dalessandro 1973).

Although evidence indicates the occurrence of three successive mitotic cycles prior to xylogenesis in tuber explants (Malawer and Phillips 1979, Phillips 1981 a), the stage of development of the parent tuber is an important factor in determining the prerequisites for differentiation. Gamma irradiation of tuber explants, of sufficient intensity to prevent DNA synthesis and mitosis, completely inhibited xylogenesis in explants from mature tubers but not immature tubers (Phillips 1981 b). The capacity for direct differentiation, that is, from parenchyma without any prior mitosis, declines as the tuber matures. Explants taken from mature and dormant tubers require a period of cell proliferation as a prerequisite for xylogenesis. Exogenous auxin alone was necessary to induce xylogenesis in explants from immature tubers, and presumably this system contains endogenous cytokinin. The cells may be in a state of "developmental plasticity" and capable of reprogramming by the exogenous auxin (Phillips 1981 b). Explants taken from fully mature tubers, on the other hand, require both auxin and cytokinin for the induction of cell division and xylogenesis (Phillips and Dodds 1977). Additional studies have shown that 3-aminobenzamide strongly inhibits xylogenesis in explants from *Helianthus tuberosus* and *Pisum sativum* without any inhibitory effects on cell division (Hawkins and Phillips 1983, Phillips and Hawkins 1985). In animal tissues it has been shown that this inhibitor blocks cytodifferentiation and not cell proliferation by suppressing the activity of a nuclear enzyme required in DNA excision repair (Farzaneh et al. 1982, A. P. Johnson and Williams 1982). Cultured plant cells were sensitive to the inhibitory effects only during a 6-h period prior to the onset of visible signs of xylogenesis. This sensitive period occurred after the final round of mitosis when the cells were in G_1–G_0, that is, after a minimum of three mitotic cycles during the first 48 h of culture (Phillips and Arnott 1983). Both 3-aminobenzamide and nicotinamide, inhibitors of ADP-ribosyl-transferase, strongly inhibited xylogenesis in *Zinnia* mesophyll cells, and had little effect on cell division (Sugiyama and Komamine 1987).

Phillips (1987) conducted experiments on the sequential requirements for hormones in cultured explants of Jersualem artichoke tuber. Mitosis commenced at about 20 h, and by 24 h about 50% of the cells had divided. By the end of 48 h the cell number had increased by four- to sixfold. The first tracheary elements appeared after 50–52 h. All experiments involved a total of 4 days of culture. The optimal hormone concentrations for xylogenesis were found to be 0.2 mg/l^{-1} 2,4-dichlorophenoxyacetic acid combined with 5.0 mg l^{-1} benzylaminopurine. When cultures were deprived of auxin for 2 days before transfer to the complete medium

(2,4-D + BAP), the cell number was reduced by 80% compared to the controls and only a trace of xylogenesis was evident. Auxin deprivation for the first 3 days completely prevented cell division and xylogenesis. Cultures grown in the absence of cytokinin for the first 2 days of culture, before transfer to the complete medium for an additional 2 days, showed no reduction in total cell number or tracheary element cell count compared with controls on the complete medium for 4 days. The presence of cytokinin during the last day produced only a threefold increase in tracheary element cell count compared with cultures grown with auxin alone for 4 days. Auxin was necessary early in the culture period and during the time of rapid cell division. Cytokinin stimulated xylogenesis when present late in the culture period, concomitant with the phase of cytodifferentiation, but not when restricted to the early period (Phillips 1987). A series of "removal" experiments was also performed, that is, explants were cultured on a complete medium and subsequently transferred to a medium containing either 2,4-D or BAP. The presence of auxin was necessary only on the first day for high levels of cell division and xylogenesis. The removal of BAP had little effect on cell number, although the tracheary element count increased linearly with the duration of exposure to cytokinin. The presence of cytokinin during the first day had little effect on xylogenesis. All of the media employed by Phillips (1987) were supplemented with rifampicin for the suppression of bacterial contamination (Phillips et al. 1981).

Watson and Halperin (1981) examined the effects of hormones and sugars on xylogenesis in cultures of Jerusalem artichoke, as well as technical problems which might account for inconsistencies between different laboratories. The alkaline maceration technique used by Minocha and Halperin (1974) probably gave abnormally high numbers of tracheary elements because of the degradation of nonxylem cells. The same study indicated that 5.0 mg l^{-1} BAP was optimal for xylogenesis, and it was later found that the optimal level was 1.0 mg l^{-1} (Watson and Halperin 1981). The effect of gibberellic acid on xylogenesis in this system has given conflicting results, and it may involve the use of suspension cultures by Halperin's group (Chap. 2.4).

A new technique has been introduced involving tuber explants of Jerusalem artichoke (Markland and Haddon 1982). The freshly prepared explants were placed on a preculture medium for 3 days prior to the induction of xylogenesis. The preculture medium contained basal salts (Gamborg et al. 1968), β-NAA (5.4 μM), BAP (22.2 μM), soluble potato starch (4% w/v), sucrose (2% w/v), thiamine-HCl (0.89 μM), and agar (1% w/v). Xylogenesis was then initiated on a similar agar medium containing 2,4-D (0.45 μM), kinetin (9.3 μM), and sucrose (3% w/v). A third mixture (nonxylogenic control) was employed in which the β-NAA of the preculture medium was replaced with α-NAA (0.54 μM). The rationale for the use of an inactive auxin isomer (β-NAA) in the preculture medium was not discussed, although there was some evidence that the hormonal mixture played some role. Subsequent culture on a xylogenic medium revealed that after preculture on a medium without any hormones, there was a marked inhibition of xylogenesis in comparison to explants receiving β-NAA. With the use of this two-step technique, up to 36% of the total number of cultured cells differentiated into tracheary elements between 1–3 days after transfer to the xylogenic medium (Markland and Haddon 1982).

Auxin-induced xylogenesis has been studied in explants prepared from the mature storage root of turnip (*Brassica campestris* cv. Rapifera) (Gersani 1987a). Discs of agar containing IAA were placed on the upper side of the explants, and the treatment induced cell division and vascular differentiation in the plane of the cambium. The auxin-induced activity was restricted to the cambium and its derivatives. Cytodifferentiation was not initiated with agar discs containing BAP, GA_3, or sucrose, respectively. Auxin-induced xylogenesis was completed within 72 h, and the timing was consistent with the induction of xylogenesis in several other systems (Gersani 1987a).

4.7 Differentiation in Suspension Cultures Including the *Zinnia* System

Aside from studies with mesophyll cells of *Zinnia*, relatively little has been done on the induction of xylogenesis in liquid cultures. With few exceptions cells suspended in a liquid medium show a reluctance to form tracheary elements. To our knowledge, no study has been undertaken on the induction of sieve element formation in suspension cultures. Single isolated cells of *Centaurea cyanus* in suspension were observed to form tracheary elements (Torrey 1975). Shake cultures of *Brachycome lineariloba* formed relatively large numbers of tracheary elements (Gould 1978). An agar-grown *Eucalyptus* callus produced tracheary elements, but the same tissue cultured in a liquid medium failed to exhibit xylogenesis (Sussex and Clutter 1967). Wilbur and Riopel (1971a, b) studied a nondifferentiating suspension of *Pelargonium* consisting of single cells and small aggregates. Reaggregation of the cells in nylon cones resulted in tracheary element formation in the vicinity of the cone apex. Vascular differentiation occurred in cellular agggregates after the aggregates had reached a certain size in wild carrot cultures (Steward et al. 1958) and in cell suspensions derived from peanut (*Arachis hypogaea*) cotyledon (Verma and van Huystee 1970). It has been suggested that cells in a liquid medium are subjected to leaching, and they may be incapable of maintaining a sufficient level of metabolites necessary for the initiation of cytodifferentiation. After the aggregates reach a sufficient size, the centrally located cells are isolated from the aqueous medium and then are capable of initiating differentiation (Steward et al. 1958). An alternative view is that the centrally located cells are subjected to mechanical forces, and that stress-induced ethylene plays a role in the induction process (Chap. 5.5, 5.6). According to Aloni (1980), sieve element differentiation must invariably precede xylogenesis, and a phloem-localized metabolite plays a role in the induction of tracheary element differentiation. This hypothesis should be tested with the *Zinnia* system consisting of isolated mesophyll cells, since relatively large numbers of these cells can be induced to form tracheary elements. Sieve element differentiation has not been examined in this system.

Fukuda and Komamine (1980b) examined the various cultural factors influencing differentiation. Optimal results were obtained with NAA (0.1 mg l^{-1}), BAP (1.0 mg l^{-1}), a low level of ammonium chloride, and an initial cell population density within the range of $0.4–3.8 \times 10^5$ mesophyll cells ml^{-1}. The osmotic potential was adjusted with mannitol, and aeration was by means of a roller apparatus. Tracheary elements first appeared 6 days after culture in cells isolated from the youngest leaves, whereas 10 days were required for the initiation of xylogenesis by mature mesophyll cells taken from older leaves. This difference in timing may be explained by the observations of Hahne and Hoffmann (1985). The cortical microtubular net, which is typically present in immature mesophyll cells, was absent or greatly reduced in mature mesophyll of *Nicotiana, Petunia*, and *Brassica* species. One of the early events during dedifferentiation of a mature mesophyll cell is the reestablishment of a network of cortical microtubules prior to spindle formation. Following the isolation of protoplasts from mature cells, division was not evident until after 78 h. Protoplasts from immature cells divided as early as 24 h after isolation (Hahne and Hoffmann 1985).

The *Zinnia* system consists of two distinct cell populations; some cells differentiate directly into tracheary elements, whereas others require some cell cycle activity. DNA replication and mitosis were timed during cytodifferentiation using a combination of microspectrophotometry, microfluorometry, tritiated thymidine autoradiography, and serial observations (Fukuda and Komamine 1981 a). Nuclei of the mesophyll cells were initially in G_1 at the 2C level of DNA. The cultured cells then underwent DNA replication at 42 h and mitosis at 54 h. The first cell cycle required a total of 58 h. More than 55% of the nuclei of the immature tracheary elements were unlabelled with tritiated thymidine and hence at the 2C level of DNA. These cells differentiated in G_1 without entering S. Approximately 30% of the tracheary elements were formed after one round of the cell cycle. The remainder differentiated following S in G_2 (Fukuda and Komamine 1981 a).

Since it would be advantageous to study xylogenesis in the absence of cell division, the inhibitory effects of nutrient limitation (low phosphate) and gamma irradiation on cell division were examined in the *Zinnia* system (Sugiyama et al. 1986). Nutrient limitation was unsuccessful in completely blocking cell division. Gamma irradiation, however, completely halted cell division without decreasing the numbers of tracheary elements formed. The irradiation delayed the appearance of secondary wall thickenings by about 20 h, although the synchrony of differentiation was unaffected (Sugiyama et al. 1986). Phillips (1981 b) also found that xylogenesis occurred in gamma-irradiated tubers of *Helianthus tuberosus* in which mitosis was completely blocked.

Lignin biosynthesis has been studied in xylogenic cultures of *Zinnia* (Fukuda and Komamine 1982 a). Lignification occurred only in a medium containing xylogenic concentrations of NAA and BAP, and not in a similar medium containing a concentration of BAP (0.001 mg l^{-1}) too low to support cytodifferentiation. Lignification occurred in a fairly synchronous manner approximately 76 h after culture, that is, about 5 h after the first visible signs of secondary wall formation in tracheary elements. The level of soluble phenolic compounds was apparently unrelated to xylogenesis. Phenylalanine ammonia-lyase (PAL; EC 4.3.1.5) activity paralleled cytodifferentiation and PAL activity was high during lignin biosynthesis (72–96 h). 0-Methyltransferase (EC 2.1.1.6) activity was higher in the non-xylogenic medium than in the inductive medium, thus indicating that this system is not a marker for xylogenesis. Other workers have obtained somewhat different results with this enzyme (Haddon and Northcote 1976 a, Kuboi and Yamada

1978a). Peroxidases (EC 1.11.1.7) associated with the cell walls peaked in activity at 72 h, just prior to lignification, and again at 84 h during active lignification. Isoperoxidase patterns were correlated with histological development in tobacco epidermal explants induced to produce either callus, vegetative buds, or floral buds (Kay and Basile 1987). There were at least three isoperoxidases that were closely correlated with lignification and xylem differentiation. The relationship between these results and the two isoperoxidases reported earlier by Fukuda and Komamine (1982a) is not known. Both PAL and wall-bound peroxidases were considered to be markers for xylogenesis by Fukuda and Komamine (1982a). Other workers have also indicated that PAL activity was closely related to xylogenesis (Rubery and Northcote 1968, Rubery and Fosket 1969, Haddon and Northcote 1976a, Chap. 5.4, 5.5). Peroxidase activity was associated with xylogenesis according to Hepler et al. (1972) and Kuboi and Yamada (1978a). Minocha and Halperin (1976), however, found little correlation between PAL and peroxidase activities with lignification and xylogenesis in Jerusalem artichoke cultures. In *Phaseolus* callus cultures peroxidase activity was negatively correlated with xylogenesis (Haddon and Northcote 1976a).

An investigation was undertaken on the isoenzyme patterns of fractionated peroxidases during xylogenesis by Masuda et al. (1983). The peroxidases were separated into five fractions: soluble, ionically wall-bound, released from wall by wall-degrading enzymes, tightly bound to the cell wall, and extracellular. Several significant changes were related to tracheary element differentiation. A new cathodic isoenzyme (ionically bound) appeared in the differentiating cells, and the activities of four preexisting isoenzymes from the same fraction greatly increased between 72–84 h of culture during the period of active xylogenesis and lignification. A new anodic isoenzyme (extracellular) appeared in the xylogenic cultures. In addition, a tightly bound peroxidase showed a rapid increase in activity during 72–84 h. The results suggest that a combination of tightly bound peroxidase, an anodic extracellular peroxidase, and five cathodic ionically bound peroxidases are possible candidates as marker proteins for tracheary element differentiation. A more specific marker seems to occur in the electrophoretic analysis of anionic peroxidases from roots of *Pisum sativum*. Fielding and Hall (1978a, b) have shown the presence of nine bands on staining the gels with diaminobenzidine as the substrate. Of these bands, only one and two are associated with the cell walls of the stele (5 mm from the root cap/root tip junction), bands three, four, five, and seven are soluble and apparently associated with the GERL system. The stelar cell wall fractions 15–20 mm from the root cap/root tip junction acquire two new bands (eight and nine), and although there has been little attempt to determine the roles of the last two isoenzymes, it is likely that at least one of them will be associated with lignification (Catesson et al. 1978). More recent cytochemical and biochemical work has indicated that syringaldazine oxidase is directly involved with xylem lignification in *Nicotiana tabacum*, and this evidence provides a better marker protein for lignification in angiosperms (Goldberg et al. 1983, 1985). Another group has suggested that ethylene may regulate lignification during xylogenesis via wall-bound peroxidase activity (A. R. Miller et al. 1985).

Fukuda and Komamine (1983) used the *Zinnia* system to examine RNA and protein syntheses as prerequisites for cytodifferentiation. Cycloheximide inhibited both xylogenesis and cell division at concentrations which also blocked the incorporation of labeled leucine into protein. Pulse treatments of the cells with cycloheximide at 12-h intervals indicated that xylogenesis was blocked to the greatest extent between 24–60 h, and this was the same time that incorporation of [^{14}C]-leucine into protein had a peak value. The inhibitory effect of actinomycin D on xylogenesis was marked between 24–60 h when incorporation of [^{14}C]-uridine into nucleic acid was at a high level. Evidently RNA and protein syntheses are prerequisites for the induction of xylem differentiation, and these syntheses occur between 24–60 h. Possibly this protein requirement is for the synthesis of tubulin (Chap. 2.3).

A sensitive immunoblotting procedure, using mouse monoclonal antibodies, was developed to measure minute amounts of plant tubulin (Fukuda and Iwata 1986). Freshly isolated *Zinnia* mesophyll cells contained low levels of tubulin, but the content increased rapidly between 24 and 48 h of culture prior to cell division and xylogenesis (Fukuda 1987). The pattern of increase closely paralleled the increase in cortical microtubules as estimated by ultrastructural analysis. The number of cortical microtubules in immature tracheary elements was estimated to be six times the level in freshly isolated cells. Between 48 and 72 h of culture, the cortical MT changed their distribution from random arrays to longitudinal and then transverse positions. The transverse arrays were clustered exclusively over the developing secondary wall ridges, which supports the earlier findings of Falconer and Seagull (1985a). Treatments to prevent cell division and xylogenesis, for example, by the addition of fluorodeoxyuridine or by removal of either auxin or cytokinin from the medium, inhibited the increase in tubulin during cell culture (Fukuda unpublished).

Wall formation and microtubule organization during xylogenesis has been followed with fluorescence microscopy and immunocytochemical techniques in *Zinnia* (Falconer and Seagull 1985a). Calcofluor white stains secondary wall material before it can be detected with either phase contrast or polarization optics. In *Zinnia* there are 3 progressive stages of development: banding of microtubules prior to secondary wall formation; clusters of microtubules subtending wall material as revealed with Calcofluor white; and finally a complex pattern of microtubules as shown with Calcofluor white, phase contrast, and polarization optics (Fig. 4.5). The cortical microtubules reorient from a longitudinal position to transverse arrays prior to the differentiation of tracheary elements. This reorientation was demonstrated by the application of taxol during the early stages of culture (Falconer and Seagull 1985b). Taxol stabilizes the microtubules in the longitudinal direction. Tracheary elements formed in taxol-treated cultures were found to possess secondary wall thickenings parallel to the long axis of the cells, whereas tracheary elements formed in control cultures invariably had transverse wall bands. A further study revealed two general types of secondary wall patterns in the *Zinnia* xylem elements: bands (scalariform) and webs (reticulate) (Falconer and Seagull 1986). Band patterns were found in elongated cells and cell regions, whereas web patterns were associated with isodiametric cells and cell regions. The position of microtubule arrays regulates both cell shape and secondary wall pattern. A microtubule depolymerizing agent amiprophos-methyl (APM) was used to induce isodiametric cell growth. Drug removal resulted in repolymerization of

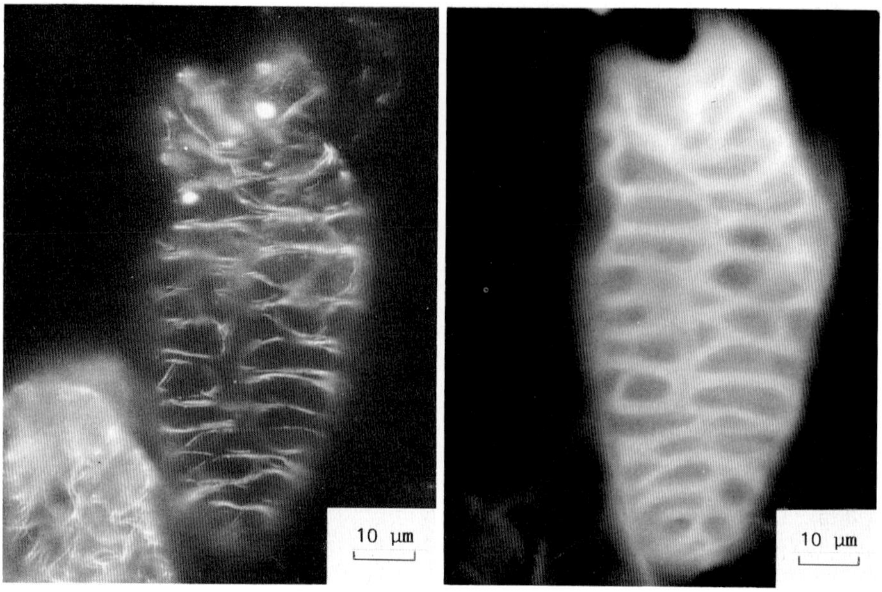

Fig. 4.5. Tracheary element differentiation by *Zinnia* mesophyll cell showing secondary wall thickenings. *Left* Arrangement of microtubules as revealed by immunocytochemical staining and fluorescence microscopy. *Right* Calcofluor white staining indicates the pattern of secondary wall thickening. Note that the arrangement of microtubules is identical to the secondary wall pattern. Scale bar = 10.0 µm. (Falconer and Seagull 1985a)

the interphase microtubules at random. Subsequent tracheary element differentiation by these spherical cells resulted in a web secondary wall pattern (Falconer and Seagull 1986).

A preliminary report by Japanese scientists indicates that actin filaments may play an important role during xylem differentiation in *Zinnia* cultures (Kobayashi et al. 1987). Double fluorescence staining of microtubules and actin filaments demonstrated the presence of aggregates of short actin filaments between microtubule bundles at the time when microtubule arrays were transverse. This phenomenon occurred immediately prior to any visible signs of secondary wall thickening as viewed with phase contrast microscopy. Rearrangement of actin filaments from longitudinal to transverse positions occurred concurrently with the shift of microtubule arrays. Disruption of actin filaments by cytochalasin B prevented the 90° shift of microtubules, and thereby resulted in the differentiation of tracheary elements with longitudinal secondary wall thickenings. The evidence suggests that the shift in position of the microtubule arrays is regulated by the actin filaments (Kobayashi et al. 1987).

Burgess and Linstead (1984) conducted an ultrastructural study of xylogenesis with *Zinnia* cells cultured in the presence of triiodobenzoic acid (TIBA). Several previous studies have been conducted on the effects of TIBA on xylogenesis (Roberts 1976). In the present study TIBA (0.5 mg l^{-1}) delayed the onset of xylogenesis by 48 h and reduced the percentage of tracheary elements from 45% to

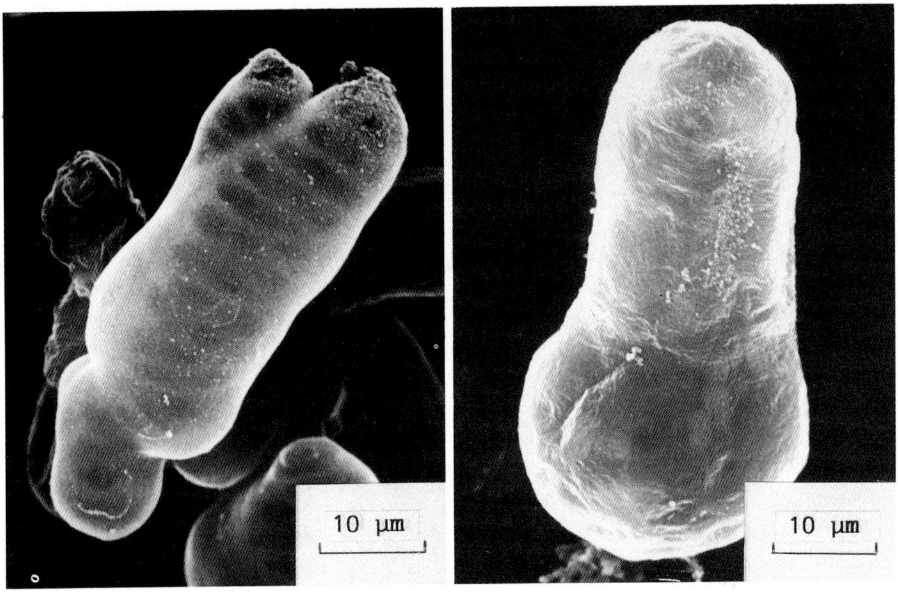

Fig. 4.6. Scanning electron micrographs of isolated *Zinnia* mesophyll cells cultured in a xylogenic medium. *Left* Cell fixed after 168 h of culture shows secondary wall thickening as weak reflecting regions. The wall at the cell tip appears porous. Scale bar = 10.0 μm. *Right* Cell fixed after 144 h of culture in a xylogenic medium containing 2,3,5-triiodobenzoic acid (5.0 mg l^{-1}). There are no secondary wall thickenings, and the cell has developed a thickened bulbous tip. Scale bar = 10.0 μm. (Burgess and Linstead 1984)

25%. A higher concentration of TIBA (5.0 mg l^{-1}) completely blocked xylogenesis over an 8-day period. Xylogenic cells inhibited with TIBA had an unusual bulbous enlargement at one end of the cells, and this abnormality may be related to the blockage of auxin flow through the cells during cytodifferentiation (Fig. 4.6). The addition of TIBA (5.0 mg l^{-1}) to cells that had been cultured for 3 days resulted in a delay of tracheary element differentiation by 2 days. Control cultures had fully developed wall thickenings on the fifth day, whereas the TIBA-treated cells showed the same stage of development on the seventh day (Burgess and Linstead 1984).

Ultrastructural observations were made on differentiating tracheary elements of *Zinnia* following a freeze fracture procedure (Haigler and R. M. Brown 1986). Rosettes composed of six subunits were found associated with the plasmalemma, Golgi cisternae, and Golgi-derived vesicles. The rosettes were found only over the developing secondary wall thickenings. The authors suggested that the Golgi apparatus plays an important part in cellulose synthesis and in the control of cell wall metabolism.

The question concerning the potential for isolated protoplasts to initiate xylogenesis was answered in the affirmative by Kohlenbach and Schöpke (1981) (Fig. 4.7). Protoplast were isolated from cotyledons and primary leaves of axenic seedlings of *Zinnia* using various combinations of Cellulase Onozuka R-10,

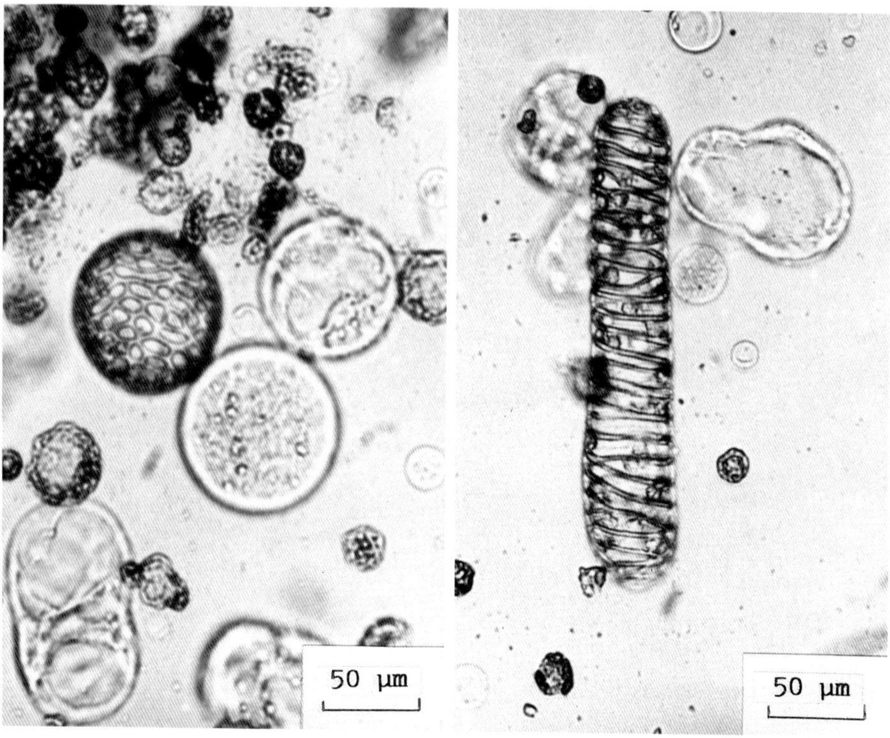

Fig. 4.7. Tracheary elements induced following a 7-day culture of isolated mesophyll protoplasts of *Zinnia elegans* on a xylogenic medium. *Left* Spherical tracheary element with web-like secondary wall pattern. Scale bar = 50 µm. *Right* Elongated tracheary element with spiral wall pattern. Scale bar = 50 µm. (Kohlenbach and Schöpke 1981)

Macerozyme R-10, and pectinase. The culture medium contained the salt mixture and vitamins given by Fukuda and Komamine (1980 b) plus casein hydrolysate, glucose monohydrate, NAA, and BAP. Tracheary elements started to appear in the cultures after 5 days. In addition, tracheary elements have been induced to form isolated protoplasts obtained from a haploid system of *Brassica napus* (Kohlenbach et al. 1982 b). The yield in the latter system can be as high as 600 tracheary elements per ml of culture.

The effect of chlorsulfuron on cell division and xylogenesis was examined in suspension cultures of *Solanum carolinense* (Reynolds 1986). This herbicide inhibits cell division by stopping cell cycle progression in G_1 and G_2 without affecting S and M phases. The chemical blocks the first step in the biosynthesis of branched amino acids, and the inhibitory effect can be reversed by the addition of isoleucine and valine to the medium (Rost and Reynolds 1985). Although chlorsulfuron almost completely blocked cell division in the *Solanum* cultures, the addition of the two amino acids restored cell proliferation. Chlorsulfuron had no discernible effect on tracheary element formation. These two amino acids presumably are required for cell cycle activity, but are unnecessary for the initiation of cytodifferentiation. These results also provide additional evidence that cell division is not a prerequisite for the initiation of xylogenesis in vitro.

4.8 Cell Wall Metabolism During Xylem Differentiation

Several contributions concerning cell wall metabolism during xylem differentiation have come from Northcotès laboratory (Northcote 1984). Secondary wall formation during xylem differentiation is characterized by several biochemical events: shutdown of pectin biosynthesis; increase in cellulose and hemicellulose production; and lignin formation. In addition, callose is synthesized during the formation of sieve elements. Two approaches have been used by Northcote and his colleagues. One method involves vascular nodule formation in suspension cultures of *Phaseolus vulgaris*, and the second employs the isolation of samples from the bark of *Acer pseudoplatanus, Populus robusta, Pinus sylvestris*, and *Abies grandis* trees. Cambial cells, immature xylem, and completely differentiated xylem elements were analyzed separately. The degree of separation into distinct cell types, that is, "purity" of the samples, may be questioned, since the different fractions were obtained by simply scraping the inner layer of bark strips (Dalessandro and Northcote 1977a).

Studies were conducted on phenylpropanoid metabolism and lignin synthesis in relation to xylem differentiation in bean suspension cultures (Haddon and Northcote 1975, 1976a, Bevan and Northcote 1979a, b, D. H. Jones and Northcote 1981). Xylogenesis occurred after 3–4 days following transfer to the NAA-kinetin induction medium, and at that time PAL activity increased to a maximum. This increased PAL activity, blocked by both actinomycin D and MDMP [D-2-(4-methyl-2,6-dinitroanilino)-*N*-methyl propionamide], provided evidence that translation and transcription were necessary for this response (D. H. Jones and Northcote 1981). The induction of PAL by the two growth regulators can be separated in time, indicating that different receptor sites may be required. NAA was required 2 days prior to the increase in PAL activity, whereas kinetin had a stimulatory effect immediately prior to the expected rise in PAL activity (Bevan and Northcote 1979b). The addition of auxin at the time of subculture changed the pattern of protein synthesis by altering the transcription pattern of mRNA, whereas kinetin did not produce this effect (Bevan and Northcote 1981a, b). Bean suspension cultures grown on a maintenance medium which does not induce cytodifferentiation gradually lose their potential to exhibit differentiation when subsequently transferred to an induction medium (Haddon and Northcote 1975, Bevan and Northcote 1979a). Both PAL activity and xylogenesis progressively declined during culture on a maintenance medium containing 2,4-D. When the maintenance medium was supplemented with coconut milk, the cells did not lose their potential to induce PAL activity and xylem differentiation. The loss of morphogenetic potential is evidently not due to cell selection or to genetic alterations, but is probably a result of reversible changes in the hormonal requirements of the cells associated with induction.

Primary wall development, associated with cell enlargement, ceases during the initiation of xylem differentiation. Shortly thereafter secondary wall formation commences. At this time there is a shift in the metabolic requirements for carbohydrates, that is, there is a greater demand for monomers leading to the production of cellulose and hemicellulose. Several alternative means of regulating this metabolic transition exist. UDP-D-Glucose dehydrogenase (EC 1.1.1.22) activity may be altered by UDP-D-xylose via feedback control (Northcote 1984). The epimerases interconnecting the hemicellulose monomers with the pectin monomers appear to be of little importance for control purposes, since tissues at all stages of differentiation contain sufficient enzyme activity to produce all the precursors of both pectin and hemicellulose (Dalessandro and Northcote 1977a). The nucleotide sugars must be transported from the cytosol across the endomembranes of the endoplasmic reticulum and the Golgi cisternae to the site of the synthase reactions for polymer formation. This transport process is doubtless an important control point for the regulation of noncellulosic polysaccharide formation (Northcote 1984). The activity of polygalacturonic acid synthase dropped sharply as sycamore cells differentiated from cambial cells to xylem, and this decreased activity was correlated with the cessation of pectin deposition (Bolwell et al. 1985). At the same time, increased xylan synthase activity was closely correlated with xylan deposition in the secondary wall thickenings of xylem cells (Dalessandro and Northcote 1981 a, b). During the induction of phloem differentiation in bean cultures there was increased callose synthase activity leading to the formation of β-1,3-glucan associated with the sieve plates (Haddon and Northcote 1975). The activities of xylan and arabinan synthases during induced cytodifferentiation required both transcription and translation processes (Bolwell and Northcote 1983b).

Particulate membrane preparations have been isolated from cambial cells, immature xylem cells, and completely differentiated xylem from the main stem of *Pinus sylvestris* trees (Dalessandro et al. 1986). These preparations synthesized a β-1,4-glucomannan using guanosine 5′-diphosphate mannose as a substrate. The membrane preparations also possessed a 2-epimerase for the conversion of GDP-mannose to GDP-glucose. The synthesis and deposition of glucomannan is a characteristic of the xylem secondary thickening in pine wood, and the activity of this glucomannan synthase (EC 2.4.1.32) increases approximately fourfold during differentiation in order to produce the additional hemicellulose (Dalessandro et al. 1986).

4.9 Some Final Thoughts on Tissue Culture Systems

The hypothesis was advanced by Wetmore and Rier (1963) that low sugar concentrations favored xylogenesis and high sugar levels enhanced phloem formation. This concept was not supported by Aloni (1980) in his study on the cultural requirements for vascular differentiation in callus cultures of *Daucus carota, Syringa vulgaris, Glycine max, Helianthus annuus, Hibiscus cannabinus*, and *Pisum sativum*. A low concentration of IAA (0.1 mg l^{-1}) induced sieve element differentiation without xylogenesis. Higher levels of IAA (0.25–1.0 mg l^{-1}) produced both xylem and phloem elements (Fig. 4.8). Varying the concentration of sucrose with a fixed concentration of IAA had no discernible effect on either phloem dif-

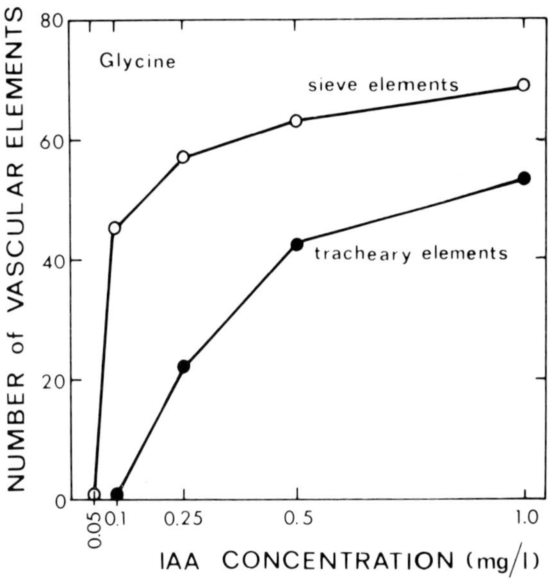

Fig. 4.8. Effect of IAA concentration on the cytodifferentiation of sieve elements and tracheary elements in tissue culture of *Glycine max* after 21 days. Sucrose concentration was 3.0% (w/v). Each *point* represents an average of three replicates from five callus tissues. (Aloni 1980)

ferentiation or on the ratio between phloem and xylem formation. These quantitative studies were made on fluctuating the levels of auxin and sugar in cultures of *Syringa, Daucus*, and *Glycine*. Phloem element differentiation commenced prior to the appearance of xylem elements in all of the cultures. Sucrose was sterilized by ultrafiltration in all of the experiments. The concentration of sucrose (1.0–6.0% w/v) was not a determining factor in controlling the type of vascular differentiation in these callus cultures. Phloem element differentiation evidently was a direct response to a very low level of auxin, whereas the induction of tracheary element differentiation required a somewhat higher concentration of auxin plus some unknown phloem-localized factor. Further work is necessary in order to establish the existence of the hypothetical xylem differentiation factor (Savidge and Wareing 1981 a).

Aside from xylem and phloem differentiation, there are few reports on the induced differentiation of other cell types in tissue cultures. Suspension cultures of *Asparagus plumosus*, however, can be induced to exhibit two distinctive types of cytodifferentiation in response to changes in the ratio of phosphate to cytokinin in the medium. A relatively high concentration of BAP (1.0 mg l^{-1}) combined with a low level of NaH_2PO_4 (0.2 mM) induced tracheary element differentiation. Cultures grown with the same low level of phosphate coupled with a low concentration of BAP (0.001–0.01 mg l^{-1}) produced a large population of storage parenchyma cells rich in amyloplasts and relatively few tracheary elements (Albinger and Beiderbeck 1983).

Vascular differentiation was examined in callus cultures of *Cryptomeria japonica* (Yamamoto et al. 1983). Xylem explants from stems were cultured for 100 days on a medium containing NAA and sucrose. The resulting callus contained a cylinder of newly formed vascular tissues adjacent to the original xylem of the explant. The xylem-like tissue had secondary wall thickenings and bordered pits, and it was separated from a phloem-like tissue by a cambial zone. In the periphery of the callus there were vascular nodules. The grafting of a lateral bud on the callus resulted in an extensive "border-like" vascular tissue beneath the bud which formed a connection with the newly formed cylindrical vascular tissues of the callus. The results were different in comparison to vascular nodule formation induced in *Syringa* following a bud graft (Wetmore and Rier 1963).

The effect of IAA on stem radial growth, tracheid lumen diameter, and tracheid wall thickness was determined in isolated stem segments of *Pinus radiata* (Sheriff 1983). Samples were removed from clones producing high density wood and from clones forming low density wood. The maximum increase in stem radius and tracheid lumen diameter occurred at IAA concentrations of 6–10 mg l^{-1}, whereas tracheid wall thickness increased in a linear manner as IAA levels increased up to 20 mg l^{-1}. Clonal differences in tracheid wall volume per unit stem volume could account for differences in wood density only within a range of 4–10 mg l^{-1} IAA.

The exogenous hormonal requirements for xylogenesis in tissue cultures may be confused because of cytokinin autotropy occurring during the culture period (Fosket 1980). This may have been a factor in the results given by Clutter (1960) and Earle (1968) (Chap. 4.2). Both reports concerned xylogenesis in cultures of freshly excised pith tissue in which xylogenesis was induced by auxin in the absence of exogenous cytokinin. Limited callus production was evident in both systems. Since cell division is regulated by a combination of auxin and cytokinin in tobacco pith (Fosket 1977), these two reports may be examples of auxin-induced xylogenesis in the presence of an endogenous cytokinin. The experiments of A. R. Miller et al. (1984) involving the induction of xylogenesis by combinations of auxin and ethylene precursors may have involved the synthesis of endogenous cytokinin (Chap. 5.5).

J. and P. M. Warren Wilson (1984) discussed the use of positional information by meristematic cells in the regulation of cytodifferentiation. One remarkable example of regeneration was described about a large wound of an actively growing dicotyledon stem. A strip was removed approximately 10–30 mm in length and about one-third of the stem diameter in depth, thus interrupting the vascular cylinder. Shortly thereafter a surface callus appeared containing two cambial zones. Derivatives of the external cambium differentiated as cork cells, and the innermost cambium produced secondary phloem and xylem in the normal positions. The regenerated vascular cambium united with the severed portions of the original vascular cambium. Other examples of the regeneration of specialized cell types include sclerenchyma found outside of the vascular cylinder, oil glands in the cortex of *Eucalyptus*, and the endodermis of certain wounded roots. It may be convenient to subdivide regeneration of vascular tissues into two categories, that is, developmental events that occur over a short time interval and long-term events. Short-term development consists of strands of tracheary elements and sieve elements formed a few days after wounding, within a rather restricted zone and showing little further development. On the other hand, long-term development refers to processes occurring during a period of 1–2 months. Long-term structures consist either of a sheet or cylinder of secondary vascular tissue produced by a regenerated vascular cambium or, in the case of tissue cultures, the formation of vascular nodules. Morphallaxis describes a remodeling of the remaining parts of an organism without the addition of any new material. In the case of plants, this would be shown by the redifferentiation of cortical tissue into vascular elements with relatively little cell division. Epimorphosis refers to extensive wound proliferation, involving callus in plants, in which differentiation occurs and reestablishes the original pattern of tissues (J. Warren Wilson and P. M. Warren Wilson 1984). The theoretical background for the development of tissue patterns is based on the presence of morphogenetic fields, that is, spatial gradients of physiological factors or morphogens (Turing 1952). The most likely morphogen for vascular differentiation is auxin. Secondary roles are indicated for cytokinin, a carbon source, and possibly ethylene. The gradient induction hypothesis states that the combination of auxin and sucrose are the two morphogens in vascular differentiation (J. Warren Wilson 1978). In recovery from large wounds, regenerating cambia appear to extend from the cut ends of the original cambia. This response appears to be an example of homeogenetic induction, that is, the influence of a differentiated cell on an adjacent cell such that it causes a similar differentiation in the adjacent cell (Blackmore and Tootill 1984). An analogous situation is seen in the extension of the interfascicular cambium from the vascular bundle into the cortex. According to J. Warren Wilson (1978), homeogenetic induction by the vascular cambium into an adjacent tissue is due to a certain ratio of auxin and sucrose. A high auxin : sucrose ratio is associated with secondary xylem, and a low ratio of the same morphogens is found in the secondary phloem. Presumably the morphogens diffuse across the cambial zone in the manner of a source-sink relationship. In a similar vein, Northcote (1985 b) suggested that the spatial arrangement of phloem, cambium, and xylem results from a feedback system involving the differentiating cells themselves. The terminal stage of xylogenesis is characterized by cell autolysis and a liberation of tryptophan, a precursor of IAA (Sheldrake and Northcote 1968 a, Sheldrake 1973). Assuming the differentiating xylem to be a source of auxin, then the amount of auxin available to the xylem should be greater than the amount present in the phloem. Also the distribution of sucrose would be different on both sides of the cambium, since it is transported in the phloem. The cambium thus represents an intermediate zone for partitioning the appropriate levels of morphogens to the two differentiating vascular tissues. If the type of tissue that differentiates is solely a function of the ratio of auxin to carbohydrate, then one should be able to selectively induce the formation of a single type of tissue in cultured pith by choosing the appropriate ratio of the two morphogens. This has not been accomplished in vitro. Aloni (1980) has demonstrated in a variety of tissue cultures that xylem and phloem differentiation are functions of auxin concentration and are independent of carbohydrate concentration.

A new concept on the relationship between basipetal auxin movement and the regulation of morphogenesis has been proposed by Zajączkowski and his colleagues (1984, Chap. 3.2). Blocks of stem tissue consisting of phloem, cambial region, and recent secondary xylem were prepared from mature forest trees and apically fed growth regulators via strips of agar. Subsequently, bioassays were run on the basipetal efflux of auxin, calibrated for equivalent amounts of IAA. The auxin collected declined with increasing length of the tissue blocks from 5–20 mm, and additional experiments indicated the presence of an oscillating pattern of auxin flux. The average wave length of the oscillations was determined to be 20 mm from species of *Larix, Picea, Acer*, and *Quercus* (Zakrzewski 1983). The wave-like patterns were greatly reduced in the presence of 2,3,5-triiodobenzoic acid and abscisic acid. Both cytokinin and gibberellic acid, however, were capable of preventing the depressing effect of abscisic acid on the wave-like pattern of the basipetal efflux of auxin. A three-dimensional model for the control of morphogenesis by auxin wave propagation has been proposed. Theoretically, the model may explain such phenomena as the seasonal variation in the width of the secondary xylem zone (earlywood vs. latewood) and the control of microfibril orientation in the successive wall layers, S_1, S_2, and S_3 (Zajączkowski et al. 1984).

Chapter 5

Physical Factors, Hormones, and Differentiation

L. W. ROBERTS

5.1 Identification of the Physical Factors Influencing Vascular Differentiation

Physical factors may alter cytodifferentiation by regulating hormone activity or availability, by influencing the carbon source required for differentiation, or by modifying some metabolic pathway associated with this developmental process (Roberts 1983). Physical conditions involved in modifying the expression of vascular element formation include temperature, water, light, gases, mechanical stress, and acidity. In most cases the observed modifications result from an environmental stress. The roles of physical factors on xylogenesis in vitro have been summarized (Roberts 1983). Denne and Dodd (1981) reviewed the environmental control of xylem formation in trees. The latter publication emphasizes the effects of climatic conditions on wood fiber diameter and wall thickness. During tree growth a change of environment has a direct effect on the rate of cell expansion, wall thickening in differentiating xylem elements, duration of cytodifferentiation phases, and indirect effects on substrate availability and growth regulator levels (Denne and Dodd 1981).

Ethylene biosynthesis is stimulated by environmental stresses such as temperature extremes, drought, waterlogging, wounding, and mechanical forces. The possibility that stress-induced ethylene biosynthesis influences vascular differentiation will be examined in several sections in this chapter. The effects of wounding on xylogenesis are discussed in Chapter 4.

5.2 Temperature

Every plant has an optimum temperature for growth (McDaniel 1982). The growth and development of cultured plant cells have an optimum temperature range of 25–30 °C, although there are some exceptions (Martin 1980).

High temperatures stimulate xylem differentiation in tissue cultures (Naik 1965, Gautheret 1966, Phillips and Dodds 1977). Lignification of the cultures may be enhanced, and Jerusalem artichoke explants produce a woody type of growth at 31 °C (Gautheret 1966). Using the same type of plant material under different cultural conditions, Phillips and Dodds (1977) found that 30 °C was the optimum temperature for callus formation and 33 °C resulted in the highest ratio of tracheary elements to undifferentiated cells. Naik (1965) studied heat-stressed cultures of *Helianthus tuberosus* and found that 43% of the total cell count had differentiated into tracheary elements after 4 weeks of incubation at 35 °C. Presumably, xylogenic cultures maintained at high temperatures are capable of producing one or more factors that stimulate the differentiation process. The establishment of metabolic pathways leading to auxin- or cytokinin-autotropy could be involved (Syōno and Furuya 1971) as well as stress-induced ethylene. High temperatures may also increase the availability of some hormone receptor involved in the initiation of cytodifferentiation.

Gautheret (1966) made some observations on the effects of low temperature on vascular differentiation in cultures of *Helianthus tuberosus*. Vascular differentiation was first observed after 4 days of culture, and a cambium appeared following 6 days (26 °C). Incubation for 3 weeks at 15 °C resulted in the differentiation of phloem elements, but not tracheary elements. The hypothesis was made that the two types of vascular tissue have different minimal temperatures for the induction of cytodifferentiation. Xylem differentiation was observed in suspension cultures of *Picea glauca* at 23.5 °C with constant light, although similar cultures did not exhibit xylogenesis with a 14/10 h photoperiod combined with temperature cycles of 23.5/12 °C (Durzan et al. 1973). The latter conditions resulted in the biosynthesis of tannins by the cultures, and this may have blocked xylogenesis (Jacquoit 1947, Chafe and Durzan 1973). Tannins bind readily to proteins (Swain 1977) and should not be considered as specific blocking agents for xylem differentiation. It is unusual that the initiation of xylem differentiation in vitro displays a sensitivity to low temperatures, whereas the same process occurs at considerably lower temperatures in the intact plant under field conditions. Seedlings of *Pinus sylvestris* form vascular elements when grown with a thermoperiod of 17.5/10 °C (Denne 1971). Let us assume that conditions exist within the intact plant that offer low temperature protection to cytodifferentiation, whereas this protective effect is absent in isolated and cultured tissues. This protective agent might be endogenous gibberellins circulating within the intact plant, since this hormone maintains the integrity of microtubules at low temperatures. The role of microtubules in xylogenesis, mainly elucidated through studies with colchicine, has been known for some time (Hardham and McCully 1982a, b). Microtubule depolymerization occurs during low temperatures. For example, after onion leaf sheath cells had been exposed to 1 °C for a period of 2.5 h microtubules were absent from 40% of the cells, and abnormal microtubules were found in the remainder of the treated cells (Mita and Shibaoka 1984a). The low temperature effects on microtubules were prevented by pretreating the cells with gibberellic acid (10^{-4}M, 48 h) prior to the low temperature stress. This is not, however, always the case. The effects of GA_3 on the arrangement and cold stability of cortical microtubules in epidermal cells of *Pisum* were examined by Akashi and Shibaoka (1987). In untreated internodes more than 50% of the epidermal cells had microtubules arranged parallel to the shoot axis. Internodal elongation is promoted by GA_3, and this regulatory effect is associated with transverse arrays of microtubules. In fact, over 60% of the epidermal cells from GA_3-treated internodes were found to have microtubules arranged at 90° to the shoot axis. Longitudinal microtubules appeared to be resistant to low temperature (0 °C, 2 h), whereas transverse microtubules in GA_3-treated cells were sensitive to the cold. A tentative hypothesis was offered by Akashi and Shibaoka (1987) in order to explain the apparent inconsistency between the effect of GA_3 on cold stability in onion leaf sheath cells (Mita and Shibaoka 1984a) compared to pea epidermal cells. The elongation rate of onion leaf sheath was not stimulated by GA_3. Possibly the cold-stabilizing effect of GA_3 is expressed only in cells whose elongation rate is not enhanced by GA_3.

Shininger (1979a) examined xylogenesis in cortical explants removed from pea roots. Low-temperature (10 °C) treatment resulted in a delay in xylem differentiation compared to the initiation of cell division in "nonxylem" cells. After xylogenesis had been initiated, the process appeared insensitive to temperatures within a 10–20 °C range, and lowering the temperature after the differentiation of tracheary elements had no apparent effect on the rate of production of these vascular elements. Xylogenesis in pea roots appears to involve an early cold-sensitive stage followed by a cold-insensitive stage according to Shininger (1979a, b). There is no indication at this time of any biochemical events associated with either of these stages.

The suppression of vascular differentiation during low temperature in late winter and early spring can be advantageous to plants in the temperate zone. Dormant flower bud primordia in many woody plants avoid freezing injury by deep supercooling (Burke and Stushnoff 1979). The possible relationship between xylem differentiation and deep supercooling was examined in eight species of *Prunus* (Ashworth 1984). Dormant buds of six species were found to deep supercool. The floral primordia of these plants contained procambial cells, but no xylem elements. At the time vascular differentiation resumed and xylem continuity with the plant was established, the capacity to deep supercool was lost. Two species of *Prunus* were incapable of supercooling and both of them contained a well-developed xylem system that presumably provided a duct for the spread of ice into the bud. The hormonal regulation of primary xylem differentiation in these two groups of *Prunus* merits further study.

5.3 Water

Both drought and waterlogging can modify vascular differentiation. One of the major effects of water potential on the level of endogenous hormones concerns ethylene biosynthesis. The presence of water per se does not pose a problem. Cells bathed in a liquid medium can proliferate and grow readily, provided that the osmotic potential is properly adjusted and gas exchange is available. Decreasing the osmotic potential with either sucrose or polyethylene glycol (PEG) stimulated xylem differentiation in callus of *Fraxinus* (Doley and Leyton 1970). The callus formed sclereids as well as tracheary elements, and there was evidence of cambial activity. Another study involved the effects of osmotic stress induced by PEG on cytodifferentiation in *Nicotiana* callus (Bornman and Huber 1979). Xylogenesis was stimulated in callus grown on media stressed with 10% and 15% PEG compared to unstressed controls. Maximal numbers of tracheary elements formed at 10% PEG coincided with a peak of phenylalanine ammonia-lyase activity (Chap. 5.4). Media stressed with 15% PEG resulted in the highest concentration of proline biosynthesis by the callus (Bornman and Huber 1979). Proline formation is a typical plant response to water stress (Parsons 1982). There is an unconfirmed report of the stimulation of auxin-induced xylogenesis by exogenous proline in stem explants of *Coleus* (Roberts and Baba 1968). Another study on PEG-induced osmotic stress gave negative results. Wright and Northcote (1973) were unable to induce vascular differentiation in *Acer pseudoplatanus* callus in the presence of various concentrations of PEG. This particular callus has never exhibited cytodifferentiation since its isolation in 1958, and the negative results were not surprising. The experiment involved combinations of sucrose (1.3–5.2% w/v) and PEG (0–16% w/v). Although the compactness of the callus was altered by the treatment, none of the cultures contained differentiated cells.

Induced xylogenesis in lettuce pith cultures was dependent on the relative amount of residual water on the lower surface of the explant in contact with the medium (D.C. Johnson and Roberts 1978). Prior to culturing the freshly prepared explants, the excess rinse water was removed by blotting with filter paper. Other explants were blotted in the same manner and then the lower surface was rehydrated with microliter amounts of water. Rehydrated explants produced greater numbers of tracheary elements than blotted explants that were not rehydrated. The addition of a thin aqueous film may improve the flow of nutrients and hormones from the medium into the tissue. The variability in tracheary cell counts common to primary explant studies may be due to slight variations in blotting the explants. Another possible factor associated with hydration is ethylene biosynthesis. The water film could create anaerobic conditions resulting in the formation of l-aminocyclopropane-1-carboxylic acid (ACC) (Yang and Hoffman 1984). Waterlogging of tomato plants resulted in blocking the aerobic conversion of ACC to ethylene by the root system. The ACC was subsequently transported to the shoot system and aerobically converted into toxic levels of ethylene (Bradford and Yang 1980). A similar anaerobic condition may exist on the lower surface of the explant, and ACC diffusing into the tissue is then converted to ethylene (Chap. 5.5, 5.6).

The effects of flooding on stem anatomy has been examined in seedlings of *Pinus halepensis* (Yamamoto et al. 1987b), *Pinus densiflora* (Yamamoto and Kozlowski 1987a), and *Thuja orientalis* (Yamamoto and Kowlowski 1987c). Soil flooding greatly altered the vascular anatomy of the stem and increased the production of ethylene. Increased thickness of the bark resulted from a proliferation of phloem parenchyma cells and the formation of intercellular spaces in the phloem. Xylem anatomy in the stressed seedlings was dependent on stem height. In the submerged stem, near the water line, the number of tracheids per radial file was greatly reduced. In the upper part of the stem a gradual increase in xylem formation was observed. Flooding of *Pinus halepensis* (Yamamoto et al. 1987b) and *Pinus densiflora* (Yamamoto and Kozlowski 1987a) induced the formation of short, thick-walled, rounded tracheids characterisitc of compression wood. These unique tracheids developed an S_3 wall layer, which is typically absent in compression wood. Xylem tracheids resembling compression wood were not observed in flooded seedlings of *Thuja orientalis* (Yamamoto and Kozlowski 1987c). Flooding also enhanced the formation of resin ducts and xylem rays, as well as the synthesis of ACC by the roots and ethylene by the shoot system. Other investigations by this research group have described the effects of ethrel treatment on stem anatomy (Yamamoto and Kozlowski 1987b, Yamamoto et al. 1987a, Chap. 5.5).

5.4 Light

The characterisitcs of radiant energy that are of importance to cytodifferentiation include intensity, wave length or spectral quality, and photoperiod (Seibert and Kadkade 1980). Light is obviously not a limiting factor for the initiation of xylem differentiation, since several in vitro studies have employed cultures incubated in the dark. Phillips and Dodds (1977) studied the effects of light on xylogenesis in explants of *Helianthus tuberosus*. During explant preparation the tissues were exposed to diffuse daylight (1 lm m^{-2}) for approximately 60 min before dark incubation. When explants were excised in the lowest level of visible green light (0.02 lm m^{-2}) and subsequently transferred to darkness, both cell and tracheary element numbers were 50% higher after 72 h compared to explants prepared in diffuse daylight. Exposure to continuous light (4.8 lm m^{-2}; universal white fluorescent) during the 72-h culture period had no discernible effect on total cell number, although the number of tracheary elements was reduced by 40% compared to dark-grown controls (Phillips and Dodds 1977). In another study, light was found to inhibit xylogenesis in *Zinnia* mesophyll cultures (Burgess and Linstead 1984).

Differences were found between Japanese cultivars of carrot in response to xylogenesis in the presence of light (Mizuno et al. 1971). Root phloem explants of the Kuroda-gosun cultivar produced tracheary elements readily on a 2,4-D medium in the absence of exogenous cytokinin during dark incubation. Since the induction of xylem differentiation requires the presence of both auxin and cytokinin, it was presumed that an endogenous cytokinin was present. A subsequent study identified the active substance as zeatin ribonucleoside (Mizuno and Komamine 1978). Similar explants from another cultivar, Hokkaidô-gosun, formed tracheids and endogenous cytokinin only in the presence of light. Previously, Linstedt and Reinert (1975) reported that carrot callus synthesizes cytokinins during culture on a 2,4-D-containing medium. Cytokinin biosynthesis in the Hokkaidô-gosun cultivar is evidently a light-dependent process.

Whether or not the phytochrome pigment system has a direct effect on the cytodifferentiation of vascular tissues is open to question. Kleiber and Mohr (1967) found that a greater number of tracheary elements were formed in the hypocotyl of *Sinapis alba* seedlings in the presence of the far-red absorbing form of phytochrome (P-730) compared to dark-grown controls. The pattern of differentiation within the bundles and the orientation of the bundles within the hypocotyl were identical in etiolated and illuminated seedlings. Unfortunately, no other reports are available on the possible effects of red and far-red light on vascular differentiation. There is one approach that should be examined. Responses stimulated by red light involve an increase in intracellular calcium (Wayne and Hepler 1985). Assuming that red light stimulates xylogenesis, this effect may result from a change in the level of intracellular calcium. Calcium is linked with auxin-mediated responses (Chap. 2.2).

Several investigators have examined the interrelationship between light-induced phenylalanine ammonia-lyase activity, lignification, and xylogenesis. The biosynthesis of lignin requires PAL (EC 4.3.1.5) activity (Hahlbrock and Grisebach 1979). Light exerts a stimulatory effect on PAL in parsley cells (Grisebach

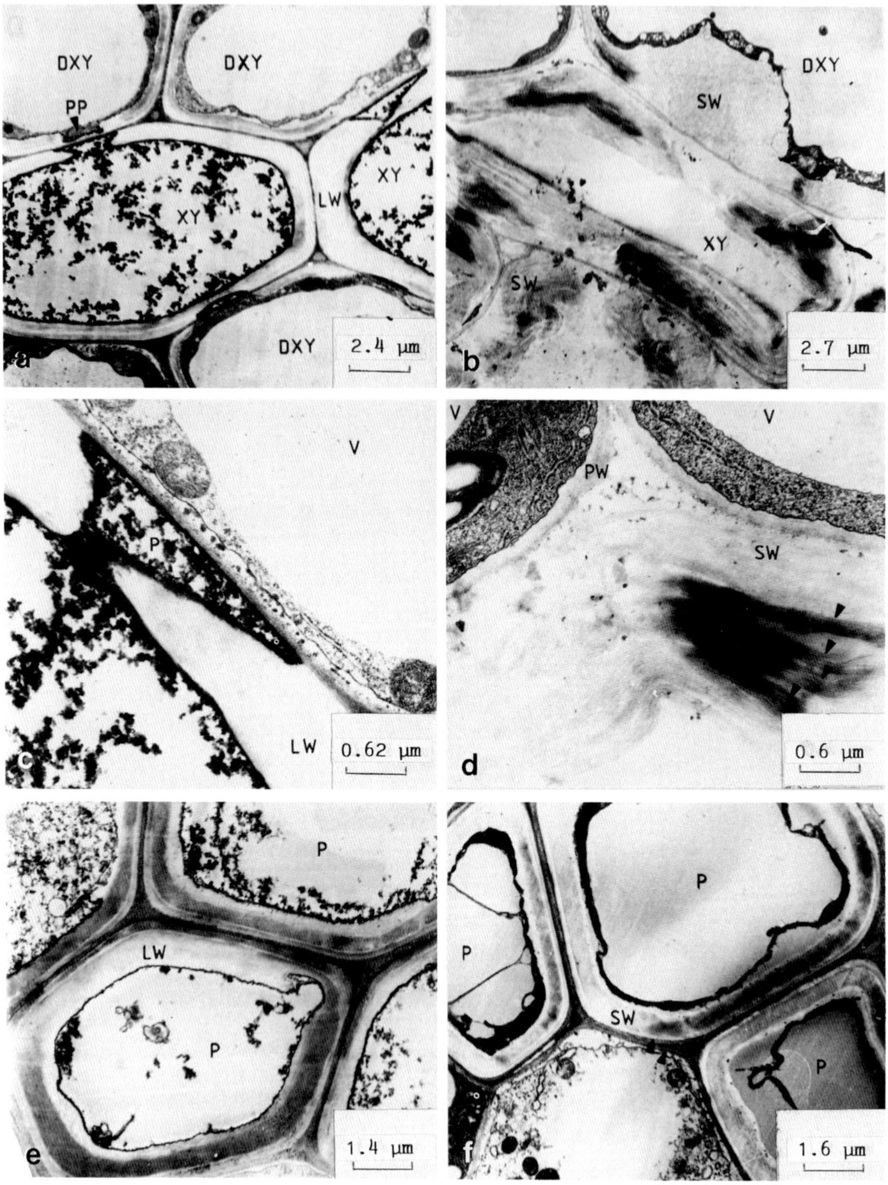

Fig. 5.1. Electron micrographs of transverse sections of the vascular tissue in the root of 7-day seedlings of *Vigna radiata* (mung bean). **a** Protoxylem of untreated tissue. *XY* mature vessel; *DXY* differentiating vessel; *PP* pit pair; *LW* lignified secondary cell wall. Darkly stained particulates are present in the mature vessels. This substance stained red with safranin/light green when viewed with the light microscope may be lignaceous. **b** Protoxylem treated with AOPP. The thicker secondary cell walls have a disordered appearance, and the mature vessel has collapsed. *XY* mature vessel; *DXY* differentiating vessel; *SW* unlignified secondary cell wall. **c** Detail of untreated protoxylem cell wall. *LW* lignified secondary cell wall, *V* vacuole of adjacent xylem parenchyma cell, *P* pit. **d** Detail of protoxylem wall following treatment with AOPP. Note separation of cellulose microfibrils. Individual fibrils are widely spaced from each other and groups of

and Hahlbrock 1974), which is thought to be due to enzyme synthesis via an increase in polysomal mRNA (Schröder et al. 1977). Light was reported also to increase PAL activity in sunflower tissue (Tena et al. 1984). Rubery and Fosket (1969) pointed out that PAL could be used as a marker for xylogenesis in *Coleus* stem explants. PAL activity and tracheary element formation peaked after 4 days of culture, and both processes declined thereafter. A close positive correlation was found between PAL activity and xylogenesis in soybean callus cultures as well as *Coleus*. Bevan and Northcote (1979a) also observed a close correlation between PAL activity and the morphogenetic potential of suspension cultures of *Phaseolus vulgaris*. Subsequently, these investigators found that the induction of PAL in bean suspension cultures was dependent on the combined presence of auxin and cytokinin. Other studies have supported the view that PAL activity is a sensitive marker for xylogenesis (Haddon and Northcote 1975, 1976a, b, Dudley and Northcote 1979). Minocha and Halperin (1976) found that PAL activity had some correlation with the amount of cytodifferentiation occurring in cultured tuber tissue of *Helianthus tuberosus*, but there were marked variations in PAL activity in tissues grown under different hormonal conditions. These fluctuations were unrelated to lignification, and they expressed the opinion that this enzyme should not be used as a marker for xylem differentiation. Fukuda and Komamine (1982a) examined lignin biosynthesis in isolated single cells of *Zinnia* mesophyll. A peak in PAL activity after 84 h of culture was coincident with the time of active lignin biosynthesis during tracheary element differentiation. Leaf mesophyll of *Zinnia* is an excellent system for biochemical studies, because the isolated cells differentiate synchronously and the percentage of cells undergoing xylogenesis reaches approximately 40% (Fukuda and Komamine 1982a). Another approach to PAL activity can be related to the stimulation of this system by ethylene (Chalutz 1973, Rhodes and Wooltorton 1973). Light, however, has been reported to have little effect on ethylene production by intact plants in an open system (Bassi and Spencer 1983). Ethylene may be more closely related to wall-bound peroxidases than to PAL during lignification of tracheary elements in lettuce pith explants (A.R. Miller et al. 1985, Chap. 5.5).

Although PAL and phenylpropanoid metabolism appear to be an early indication of xylem differentiation, it does not mean that these processes are a prerequisite for the initiation of cytodifferentiation (Smart and Amrhein 1985, Fig. 5.1). Seedlings of mung bean were treated with 1-α-aminooxy-β-phenylpropionic acid (AOPP). This compound is a potent inhibitor of PAL activity, and it serves to block the biosynthesis of both anthocyanin and lignin. The treated seedlings contained unlignified and collapsed xylem vessels displaying aberrant secondary

fibrils (densely stained) appear to be detached from other wall components (indicated by *arrowhead*). *PW* primary wall of xylem parenchyma cell; *V* vacuole; *SW* unlignified secondary wall of protoxylem element. **e** Untreated phloem fibers. *LW* lignified cell wall; *P* phloem fiber. **f** Phloem fibers following treatment with AOPP. Unlike the protoxylem vessels that were treated with AOPP, the unlignified phloem fibers do not collapse. The secondary wall strcture of phloem fibers in lignified and nonlignified tissues appear to be very similar in appearance (cf., *e* and *f*). *SW* unlignified secondary wall; *P* phloem fiber; *arrowhead* primary wall. (Smart and Amrhein 1985) walls (Fig. 5.1 b). Secondary wall deposition had occurred in the xylem cells, although the cellulose microfibrils were separated and disorganized within the lumens of the xylem elements (Fig. 5.1 d). The walls of the unlignified phloem fibers were capable of withstanding the growth stresses and did not collapse in the presence of AOPP (Fig. 5.1 f). This study provides evidence that the lack of lignification has no feedback effect on the regulation of gene expression associated with secondary wall formation.

The question of whether or not a close relationship exists between xylogenesis and lignification was examined in callus cultures of soybean (*Glycine max* L. var. Wayne) by A. R. Miller and Roberts (1986). Both processes have similar hormonal requirements and follow similar time courses, but there was no statistical relationship between tracheary cell counts and lignin content in the soybean cultures. The biosynthesis of lignin not associated with tracheary element cell walls may have been a factor. For example, polyphenol accumulation occurred in the absence of tracheary element differentiation in *Populus trichocarpa* callus (Hoffman et al. 1985). These workers also demonstrated with this system that callus containing large numbers of tracheary elements does not exhibit any more lignin than callus cultures lacking tracheary elements.

Light induced phloem anastomoses in the hypocotyl of *Cucumis sativus* L. seedlings (Aloni et al. 1986). The phloem of seedlings grown in the dark consisted only of longitudinal strands of sieve tubes in the region between the cotyledons and the root system. Illumination from cool-white fluorescent tubes induced the differentiation of lateral sieve tubes between the longitudinal strands. The number and density of phloem anastomoses increased with an increase in light intensity, according to Aloni et al. (1986).

5.5 Gases

Gases that may have an effect on cytodifferentiation during the culture of plant tissues include oxygen, carbon dioxide, ozone, and ethylene. Plants growing under field conditions are subjected to these gases, as well as to a host of air pollutants (Heath 1980, Treshow 1984). Aside from investigations on ethylene, little research has been done on the influence of gases on vascular differentiation.

The relative concentrations of dissolved oxygen in liquid suspension cultures display a regulatory effect on the type of differentiation occurring in carrot cultures (Kessell and Carr 1972). Embryogenesis was initiated by low levels of oxygen, whereas higher concentrations resulted in the initiation of roots. The oxygen concentration could influence vascular differentiation indirectly via ethylene. When oxygen was lowered to 5%, there was a reduction in the ethylene-induced suppression of growth of etiolated pea seedlings (Beyer 1979). Beyer found that oxygen (5%) inhibited the incorporation of ethylene into the tissue, and blocked the oxidation of ethylene to carbon dioxide. The conversion of ACC to ethylene is an oxygen-dependent step, and thus anaerobic conditions will deplete the supply of endogenous ethylene.

Aside from photoautotropic cell cultures grown in the light without an external supply of sugar (Yamada et al. 1982), carbon dioxide fixation is not a significant factor in plant tissue cultures in the presence of sufficient levels of organic carbon (S. M. Martin 1980). Carbon dioxide (11%) stimulated xylogenesis in cultured peach mesocarp (Bradley and Dahmen 1971), and this response may be due to fluctuations in the level of endogenous ethylene. Since CO_2 markedly stimulates ethylene biosynthesis without altering the endogenous level of ACC, presumably this effect is exerted on the conversion of ACC to ethylene (Kao and Yang 1982). The latter workers have proposed that CO_2 activates the enzyme converting ACC to ethylene. Since light or CO_2 can promote or block ethylene production, depending on the tissue system, the realtionship is indeed complex (Yang and Hoffman 1984).

Ozone suppressed xylogenesis in wounded internodes of *Coleus* and in *Parthenocissus* callus (Rier 1976). Ozone is produced during ultraviolet irradiation from germicidal lamps used in transfer chambers (Dodds and Roberts 1985). Assuming the chamber is well ventilated, a brief exposure of the tissues to traces of this gas are inconsequential. Ozone is an atmospheric pollutant, and plants exposed under field conditions can be seriously affected (Heath 1980, Treshow 1984).

Ethylene may be a hidden hormone in the initiation of tracheary element differentiation (Roberts 1976). Experimental evidence in support of this hypothesis has been indirect and circumstantial from tissue culture studies. The role of mechanical stress in the modification of secondary xylem formation, that is, reaction wood production, is discussed in Chap. 5.6. The initial efforts of Zobel and Roberts (1978) to demonstrate a role for ethylene during the induction of xylogenesis in vitro were unsuccessful, because of technical difficulties in maintaining nanoliter levels of endogenous ethylene within the tissues during the culture period. Earlier Zobel (1973) had shown that nanoliter concentrations of ethylene were capable of altering the differentiation of secondary xylem in the diagravitropic mutant of tomato. The biosynthesis of ethylene in higher plant tissues generally follows the pathway: L-methionine → S-adenosylmethionine (SAM) → 1-aminocyclopropane-1-carboxylic acid (ACC) → ethylene (Yang and Hoffman 1984). The addition of trace amounts of L-methionine (0.025 μM) to a xylogenic medium resulted in considerably greater numbers of tracheary elements in cultured lettuce pith explants compared to similar explants grown on a medium lacking this amino acid (Roberts and Baba 1978). It was assumed that the utilization of L-methionine in ethylene production was responsible for the stimulation of xylogenesis. C. Nitsch (1968) had demonstrated that a medium supplemented with methionine gave an ethylene-type response in a plant tissue culture. A. R. Miller and Roberts (1982) conducted experiments with the Clark 63 cultivar of soybean. This cultivar possesses a temperature-dependent metabolic pathway for ethylene biosynthesis. The lack of normal hypocotyl elongation of Clark 63 seedlings grown at 25 °C results from the excessive production of ethylene at this temperature (Samimy and LaMotte 1976, Samimy 1978). The seedlings grew normally at either 20 or 30 °C, and ethylene production was considerably less at these temperatures compared to 25 °C. Callus cultures were initiated from cotyledonary explants prepared from Clark 63 seeds with the assumption that the cultured cells would also demonstrate a temperature-dependent pathway for ethylene production. The hypothesis was that the increased synthesis of ethylene would synergize auxin-induced xylogenesis in the callus. The results supported the hypothesis. Clark 63 callus produced greater numbers of tracheary elements on a xylogenic medium containing L-methionine at 25 °C than at 20 °C (A. R. Miller and Roberts 1982). Oddly enough, Zobel (1987) reported that Clark 63 callus, derived from hypocotyl explants, evolved significantly less ethylene at 25 °C in comparison to callus cultures of Wayne and Chippewa. Although this does not support the A. R. Miller and Roberts (1982) hypothesis, a logical experiment would be to measure and compare ethylene output at 25 °C from callus cultures derived from both hypocotyl and cotyledon. Also, because of the possibility of intracellular binding (compartmentalization), the ambient ethylene measured by Zobel (1987) may not be an accurate representation of the total ethylene biosynthesis of the system.

Another study examined the effects of ethylene promoters and inhibitors on the induction of xylogenesis in lettuce pith cultures (A. R. Miller and Roberts 1984). The precursors L-methionine, S-adenosylmethionine (SAM), and 1-aminocyclopropane-1-carboxylic acid (ACC) promoted xylogenesis and ethylene production. Similar results were found in cultures that received the ethylene-releasing agent 2-chloroethylphosphonic acid (CEPA). The ethylene blockers aminoethoxyvinylglycine (AVG), cobalt, and silver (Yu and Yang 1979, Yoshii et al. 1980) inhibited xylogenesis. The inhibitory effects were reversed, to some extent, by the addition of exogenous ethylene or ethylene precursors.

There is evidence that ethylene may substitute for cytokinin during auxin-induced xylogenesis (A. R. Miller et al. 1984). Tracheary elements were found in explants treated with combinations of IAA-CEPA and IAA-ACC after 21 days, although no cytodifferentiation occurred in the presence of IAA, kinetin, ACC, or CEPA alone, nor with combinations of kinetin-ACC or kinetin CEPA (Chap. 2.3, Tables 5.1, 5.2).

The relationship between lignification and xylogenesis was examined in cultured *Lactuca* explants (A. R. Miller et al. 1985). Lignin production occurred in two distinct periods, one preceding xylem differentiation (days 0–3 of culture), and a second phase coincident with maximal xylogenesis (days 4–7 of culture). Soluble phenolics increased concomitant with the first phase of lignification, and then decreased during the second phase. The addition of silver nitrate to the medium inhibited only the second phase of lignification and markedly decreased wall-bound peroxidase activity. The addition of L-methionine to the medium completely reversed the inhibitory effect of silver on both lignification and wall-bound peroxidase activity. Although silver increased PAL activity, it had no discernible effect on either soluble phenolic production or soluble peroxidase activity. The data indicate that ethylene plays a role in regulating lignification during xylogenesis via wall-bound peroxidase activity. Other workers have suggested that wall-bound peroxidase is a marker enzyme for lignin biosynthesis and xylogenesis (Fukuda and Komamine 1982 a, Higuchi 1985). Ethylene enhances the activities of specific peroxidase isoenzymes (Matsuno and Uritani 1972, Morgan and Fowler 1972).

The vascular anatomy of seedlings of *Ulmus americana* (Yamamoto et al. 1987 a) and *Pinus halepensis* (Yamamoto and Kozlowski 1987 b) were greatly altered following treatment with ethrel (2-chloroethylphosphonic acid). The ethyl-

Table 5.1. Effects of various hormone regimes on ethylene production and xylogenesis in *Lactuca* pith explants. (A. R. Miller et al. 1984)

Treatment[a]	Ethylene production at following days in culture				Fresh wt[b] mg/explant	Xylem elements[b] no./g fresh wt
	1	7	14	21		
	nl/h/explant					
No hormones	0	0.2±0.0	0.2±0.0	0.2±0.0	70± 2	0
IAA	0	1.0±0.0	0	0	109± 3	0
Kinetin	0	0.1±0.09	0.5±0.03	0.6±0.03	154± 9	0
ACC	0.8±0.1	2.4±0.06	1.5±0.03	0.5±0.0	73± 3	0
CEPA	3.6±0.03	2.0±0.03	2.2±0.03	1.1±0.09	79± 2	0
IAA+kinetin	0.4±0.06	11.7±0.03	11.8±0.3	18.0±0.5	364±30	250000±30000
IAA+ACC	0.6±0.06	6.0±0.8	9.4±0.6	5.5±0.7	100± 6	1700± 600
IAA+CEPA	4.0±0.1	1.8±0.09	1.9±0.06	0.9±0.03	108± 6	3900± 600
Kinetin+ACC	3.6±0.06	0.8±0.03	0.7±0.0	1.0±0.06	88± 3	0
Kinetin+CEPA	2.2±0.06	2.5±0.09	1.8±0.06	0.8±0.03	90± 6	0

[a] Explants were cultured on basal medium supplemented as specified with IAA (60 µM), kinetin (0.5 µM), ACC (10 µM), and CEPA (1 µM).
[b] Determined after 21 days. Initial fresh weight was approximately 52 mg.
[c] Values reported are from a representative experiment; each treatment was repeated three times. Values expressed ± SE ($n=10$).

Table 5.2. Time course of IAA-ACC-induced xylogenesis in *Lactuca* pith explants. (A. R. Miller et al. 1984)

Treatment[a]	Xylem elements at following days in culture		
	7	14	21
	no./g fresh wt		
IAA	0	0	0
IAA + 0.1 μM ACC	0	0	0
IAA + 1.0 μM ACC	0	50 ± 20	9000 ± 1600
IAA + 10 μM ACC	0	300 ± 75	28000 ± 1000
IAA + kinetin	55000 ± 2100[b]	188000 ± 31300	344000 ± 62100

[a] Explants were cultured on basal medium supplemented as specified with IAA (60 μM), kinetin (0.5 μM), and ACC.
[b] Values expressed ± SE ($n = 8$).

ene-releasing agent was applied in lanolin paste at various concentrations to the stems. Although the plant responses varied with ethrel concentration, the following changes were noted in the *Ulmus* treated seedlings: (1) increased phloem and xylem formation; (2) increased production of vessels with decreased vessel diameter; (3) abnormally large xylem ray cells and increased number of cells per xylem ray; (4) the production of longitudinal resin ducts in the xylem; and (5) the development of dark organic deposits in vascular parenchyma cells and immature fibers. The organic deposits were consistent with previous reports on the enhanced production of phenolic compounds by ethrel (Hillis 1975). The evidence that ethylene is a hormone in the initiation of xylogenesis has been discussed by Roberts and A. R. Miller (1982).

5.6 Mechanical Stress

Ethylene biosynthesis is readily stimulated in response to external pressure or tension, vibration, wind or physical contact (Jaffe 1980). These stimuli result in developmental changes in growth known collectively as thigmomorphogenesis (Jaffe 1973). Mechanical perturbation of bean plants produced such anatomical changes as decreased cell elongation in the outer stem tissues, increased diameter of the pith parenchyma cells, and increased production of secondary xylem. The application of exogenous ethylene reproduced all of these anatomical effects of mechanical perturbation (Jaffe and Biro 1979, Biro et al. 1980). Mechanically perturbed bean plants showed increased radial expansion of cortical parenchyma and increased production of secondary xylem. These responses were found within a few hours after a single mechanical perturbation, and treatment of the plants with the ethylene-releasing CEPA duplicated all of these anatomical effects (Biro and Jaffe 1984). Jaffe (1980) has suggested a tentative hypothesis concerning thigmomorphogenesis in bean plants. Mechanical stimuli trigger a membrane alteration that results in a burst of ethylene biosynthesis. The released ethylene, in some manner, blocks the basipetal flow of auxin. Accumulated auxin enhances cambial activity and xylem production and, at the same time, stimulates further ethylene biosynthesis. Another factor may be the increase in abscisic acid within the perturbed system (Erner and Jaffe 1982). From our limited knowledge of ABA metabolism, there is no obvious relationship between the ethylene-auxin hypothesis and the appearance of ABA (Milborrow 1983). Both mechanical perturbation and ethylene alter certain components of the endomembrane-associated proteins, phospholipids, and fatty acids. Jaffe (1980) considers that these changes in the endomembranes are related to the anatomical expression of thigmomorphogenesis.

Aside from the responses of bean seedlings to mechanical perturbation, a somewhat different system was found in *Bryonia dioica* (Boyer et al. 1979 b). Mechanical rubbing of these plants does not result in the accumulation of auxin in the lower internodes, but there is an increase in peroxidase activity induced by the stress-released ethylene. These basic (cathodic) peroxidases were assumed to play a role in IAA oxidase activity. Pretreatment of young *Bryonia* plants with lithium prevented the inhibition of internodal elongation, suppressed the appearance of a specific isoperoxidase characteristic of mechanical perturbation, and counteracted the rubbing-induced ethylene production and the capacity of the tissues to convert ACC to ethylene (Boyer et al. 1979 a, 1983). A subsequent study on the inhibitory effects of cobalt on thigmomorphogenesis led to the discovery of two different, sequentially ordered mechanisms for ACC-ethylene conversion (Boyer et al. 1986). Additional experiments were performed in order to elucidate these two conversion mechanisms (De Jaegher et al. 1987). Rubbing *Bryonia* internodes induced a transient burst of ethylene after a brief lag of 15–20 min. The release of ethylene was accompanied by enhanced ACC synthase activity, free ACC, and ACC oxidation. Also, there was a gradual rise in conjugated ACC during the first hour. A second gradual increase in ACC conversion was observed to occur between 3 and 24 h (De Jaegher et al. 1987). The addition of cobalt to the system had no discernible effect on the initial burst of ethylene, but the second rise in ACC conversion was completely blocked (Boyer et al. 1986). Rubbing-induced lignification in *Bryonia* was strongly inhibited by the presence of *N*(O-hydroxyphenol) and *N*(O-aminophenyl) sulfinamoyl tertiobutyl acetate, specific inhibitors of cinnamyl alcohol dehydrogenase (De Jaegher and Boyer 1987). The latter enzyme is strictly associated with the biosynthesis of lignin monomers. In addition to the inhibition of lignification, these blocking agents also prevented the inhibition of elongation due to the rubbing stimulus. The data support the concept that there is a close relationship between lignification and elongation in *Bryonia* internodes.

The role of pressure in cytodifferentiation was first demonstrated by C. L. Brown and Sax (1962). Bark strips of *Populus trichocarpa* and *Pinus strobus*, separated from the bole wood in the early spring, were wrapped with plastic to prevent desiccation. Undifferentiated callus was formed along the cambial zone of the bark strips. Bark strips prepared in a similar manner and pressed against the tree differentiated xylem and phloem elements from cambial derivatives. C. L. Brown (1964) described the effects of external pressure on xylem differentiation in cultured explants of *Populus deltoides*. Although Brown did not determine ethylene production by the stressed tissues, a study by Goeschl's group (Goeschl et al. 1966) demonstrated that ethylene production by physical stress in excised pea epicotyls was associated with stress-induced anatomical changes. The effects of external pressure on cytodifferentiation was observed by Lintilhac and Vesecky (1984) following a 5-day growth of *Nicotiana* pith explants in a restraining device. The compressive stress resulted in the production of radially aligned files of cells, a cambial-like zone, and evidence of xylem differentiation. Mechanically stressed branches of *Pinus strobus, Pyrus malus*, and *Prunus perisca* trees produced more secondary xylem than nonstressed control branches (Leopold 1972, K. M. Brown and Leopold 1973). The bending of apple seedlings in order to align the shoot apex in a horizontal position resulted in increased ethylene concentrations in the stressed shoots (Robitaille and Leopold 1974, Robitaille 1975). Also, the application of CEPA to shoots of pine (K. M. Brown and Leopold 1973) and apple (Robitaille and Leopold 1974) produced localized increases in stem diameter due to the production of secondary xylem. Although shaking produced thigmomorphogenic responses in *Pinus resinosa*, a greater effect was produced by twisting or applying torque stress (Quirk et al. 1975). Tendrils of *Passiflora caerulea* display thigmomorphogenic responses, and contact stimuli induced lignification of the tracheary elements within the tendrils (Reinhold et al. 1972).

Considerable interest has been shown in the hormonal regulation of reaction wood formation. Deviation of the shoot from the vertical induces the formation of tension wood in angiosperms and compression wood in gymnosperms (Zimmermann and C. L. Brown 1971, Timell 1986). Briefly, the tension wood of angiosperms is characterized by vessels with reduced width and in smaller numbers, and the fibers have a thick gelatinous layer interior to the S layers (Esau 1977). Compression wood of gymnosperms typically contains short tracheids with cell walls appearing rounded in transection. These tracheids are heavily lignified, and the S_3 layer is missing. Timell (1986) has given a detailed description of the anatomy of compression wood. Subsequent volumes in the Springer Series will be devoted to wood anatomy. Both auxin and ethylene appear to play roles in reaction wood formation, although the roles are obscure. The evidence is meager that gibberellins, cytokinins, and abscisic acid are involved (see Timell 1986).

A deficiency of auxin is associated with tension wood production, whereas an optimal level of auxin results in earlywood. A supraoptimal concentration of auxin causes the formation of compression wood in gymnosperms (Timell 1986). A positive correlation was reported between compression wood formation and the accumulation of IAA in cuttings of Douglas fir stems (Starbuck and Roberts 1983, Starbuck and Phelps 1986). Compression wood has been induced by 1-N-naphthylphthalamic acid, a transport inhibitor of IAA (Yamaguchi et al. 1980), and compression wood formation may be a general characteristic of agents effective in blocking auxin transport in woody plants. Compression wood results following the accumulation of auxin above a girdling barrier and in the presence of TIBA and morphactins (see Timell 1986). Morphactin IT 3456 induced a type of secondary xylem similar to compression wood in two coniferous species (Phelps et al. 1974). The same research group compared the anatomical effects of morphactin and CEPA on wood formation in *Salix fragilis* (Phelps et al. 1980). Another group applied morphactin in a band around the internodes of artifically inclined 5-year-old Japanese larch trees (Yamaguchi et al. 1983). Compression wood was formed completely around the inclined stems above the treated zone. On the underside of the stems below the treated region, there was an inhibition of natural compression wood formation. Presumably there was insufficient auxin below the treated zone for the induction of compression wood. The presence of a vascular cambium is evidently a prerequisite for compression wood formation. When young spruce (*Picea* sp.) trees were inclined at 45°, there was no appreciable difference in the development of the procambium-cambium continuum between the upper side and the under side of the stressed shoot (Yoshizawa et al. 1986). The formation and anatomy of the primary tracheary elements were the same, regardless of the site of the procambium in the shoot. Compression wood cells were not present until the development of the vascular cambium cylinder had been completed. The assumption was made by the investigators that the stimulus for compression wood formation was transmitted to the differentiating secondary xylem tissue only by a cambium with cylindrical continuity and/or by postcambial differentiating tracheids.

There is indirect evidence of a positive correlation between ethylene biosynthesis and reaction wood formation. Tension wood formation was induced in *Eucalyptus gomphocephala* by placing the seedlings in a horizontal position (Nelson and Hillis 1978c). The stressed plants had 60–80% tension wood by volume in the upper halves of the shoots and 0–10% in the lower halves. The upper halves had greater levels of ethylene compared to lower halves and vertical seedlings. Compression wood typically forms only on the gravitationally lower sides of leaning conifer stems and may involve ethylene production. Endogenous ACC, the immediate precursor to ethylene, was identified in the vascular cambium on the lower side of *Pinus contorta* stems in association with compression wood formation (Savidge et al. 1983). A video image analyzer was used by Telewski et al. (1983) to study xylem formation in CEPA-treated seedlings of loblolly pine (*Pinus taeda* L.). Compared to untreated controls, the tracheid lumens of the treated plants were shorter and had a smaller cross-sectional area. The ethylene treatment increased resin canal formation in the xylem, and the tracheid cell walls were thicker in comparison to the controls. Endogenous ethylene is believed to be responsible for heartwood formation in woody plants (Shain and Hillis 1973, Hillis 1975, Nelson 1978). The production of ethylene by the living cells of the secondary xylem is related to the biosynthesis of phenolic compounds produced during heartwood formation (Craker and Wetherbee 1973, Rhodes and Wooltorton 1973, Hillis 1975). Studies have been conducted also on ethylene production in the sapwood of black walnut in the early stages of heartwood formation (Nelson et al. 1981, Phelps et al. 1983). Ethylene formation in the outer sapwood and cambium may be a factor in the regulation of cambial dormancy in stands of *Eucalyptus pauciflora* growing at different altitudes (Nelson and Hillis 1978b).

There are some interesting lines of evidence concerning the role of ethylene in the metabolism of phenolic compounds. Kino is an aqueous solution of polyphenolic substances located in veins in the xylem of *Eucalyptus* (Hillis 1972, 1987). Kino vein formation in several species of *Eucalyptus* was associated with a doubling of the rate of ethylene biosynthesis, and, in addition, it was found that applications of CEPA to the cambial region stimulated kino exudation (Hillis 1975).

The evidence suggests that stress-induced ethylene in the cambial region is responsible for the kino formation (Nelson and Hillis 1978a). A related phenomenon is the biosynthesis and exudation of gum commonly found in *Prunus* species. The gum produced in sour cherry is a slightly acidic arabinogalactan (F. Smith and Montgomery 1959). Gum, produced in the immature secondary xylem and periderm, is often associated with the presence of polyphenolic compounds (Talboys 1968). Gum production is an ethylene-mediated phenomenon, and gummosis in sour cherry can be stimulated by spraying the trees with CEPA (Olien and Bukovac 1982).

According to Fisher (1985) tension wood formation in a tropical species of *Terminalia* involves two environmental factors which may or may not be additive. Gravitational reorientation of the shoot axis from its normal position is the dominant factor, and this realignment produces tension wood in such a manner as to bring the axis back to its original position. Changing the direction of the axis with respect to gravity presumably alters the level of auxin and possibly other hormones within the axis adjacent to the vascular cambium. Mechanical stress evidently has a similar effect on the endogenous hormones, although this stress cannot reverse that of gravitational reorientation when the two forces are in opposition.

Wind-induced changes in xylem anatomy of *Abies fraseri* (Telewski and Jaffe 1986b) and *Pinus taeda* (Telewski and Jaffe 1986b) have been described. A further study by these workers (1986c) confirmed the role of ethylene in the physiological responses of conifers to mechanical perturbation.

Mechanical perturbation of bean stems induced the rapid deposition of callose, both at sieve plates and elsewhere (Jaffe et al. 1985). Treatment of the bean seedlings with a spray of 2-deoxy-D-glucose, an inhibitor of protein glycosylation, blocked both stem thickening and callose deposition. The inhibitor also appeared to block ethylene production, which occurred after callose formation.

5.7 Acidity

Although there are some exceptions, most tissue culture media are adjusted to pH 5.5–5.8 before autoclaving. Media vary in regard to their buffering capacity, and the pH usually drifts during a tissue culture experiment (S. M. Martin 1980). Shiraishi (personal communication) examined xylogenesis in lettuce pith explants cultured on filter paper platforms (Dodds and Roberts 1985) impregnated with Murashige and Skoog's (1962) basal mixture. The initial pH was adjusted to 5.5, and after 7 days of culture the pH had risen to 6.8. Xylogenesis was enhanced by transferring the explants to a fresh medium (pH 5.5) after 4 days of culture. The initial pH of the medium had a regulatory effect on the type of cytodifferentiation that occurred in cultured fruit vesicles of *Citrus limon* (L.) Burmann var. Assam lemon. The cultures were grown on a xylogenic medium and subsequently produced sclereids, xylem fibers, and tracheary elements. The greatest number of sclereids and tracheary elements were found in the callus grown on a medium with an initial pH between 5 and 6, whereas pH 7 was optimal for the formation of xylem fibers. Neither callus formation nor any form of cytodifferentiation was observed at pH values below 3 (Khan et al. 1986). This research group also examined the cultural requirements for the induction of xylogenesis in isolated fruit vesicles from Assam lemon (Kulshrestha et al. 1982). Evidently the isolated vesicles contained endogenous cytokinin, since auxin was the only exogenous hormone required for tracheary element differentiation. The addition of a relatively high concentration of citric acid (2000 mg l^{-1}) substituted for auxin in the induction of xylogenesis. This is the first report that citric acid may have a xylogenic effect similar to auxin. Xylogenesis was confined to a discrete zone in the upper end of the vesicle stalk adjacent to the sac. Tracheary element differentiation in cultured *Citrus* juice vesicles has been observed by other groups (Kordan 1964, Kato 1980, Macchia personal communication).

Media pH was a factor in the regulation of morphogenetic pathways in thin cell-layer explants of tobacco (Tran Thanh Van et al. 1985). Various cytodifferentiation responses were obtained with different concentrations of kinetin, IAA, and oligosaccharin mixtures at pH values of 3.8, 5.0, and 6.0. The term oligosaccharins refers to oligosaccharides capable of regulating physiological responses (Albersheim et al. 1983). The oligosaccharins used in these experiments were obtained following the enzymatic treatment of cell walls of *Acer pseudoplatanus* suspension cultures. A variety of morphogenetic responses were induced simply by altering the pH. For example, in the absence of oligosaccharins a total of 1800 vegetative buds were produced on 20 tobacco explants at pH 6.0, whereas at pH 5.0 under the same conditions there was callus formation but no evidence of morphogenesis. In the presence of oligosaccharins, the same number of explants produced 600 flower buds and 18 vegetative buds at pH 6.0. Under the same cultural conditions at pH 5.0, there were 400 vegetative buds, but no differentiation of flower buds.

Epilogue

Xylogenesis consists of a cytodifferentiation sequence (Roberts 1976) that begins with cell determination and ends with autolysis of the newly formed tracheary element. Although the various cytological, cytochemical, physiological, and biochemical aspects of the sequence have been described, the overall process remains an incomplete puzzle. Our information is fragmentary, and the interrelationships linking the events are completely unknown.

Evidently we must examine two systems, namely, the initiation of vascular tissue in the developing embryo and the extension of the programming for xylogenesis during the growth of the mature plant. The initiation of the pre-procambium, during embryonic development, is probably under the control of hormones from the parent plant. Although cytological evidence does not permit the early detection of pre-procambial cells until late in development (King and Heyes 1986), cytochemical markers permit the early identification of these cells within 3–4 days following in vitro fertilization in *Pisum sativum* (Gahan and Rana unpublished observations) and within 5 days in somatic embryos of *Daucus carota* (Caligo et al. 1986). This points the way for determining the initial events in programming, for vascular tissue in general and xylogenesis in particular, using both in vivo and in vitro systems. The application of quantitative cytochemistry coupled with recently developed techniques of molecular biology should enable rapid progress to be made in this area.

The earliest cytochemical marker for commitment to vascular element differentiation is carboxylesterase (EC 3.1.1.1), which is active in meristematic cells determined to differentiate as vascular elements (Gahan 1981, Rana and Gahan 1982, 1983b, Gahan and Bellani 1984). Exploitation of this marker of cell determination in root wounding experiments has permitted the identification of the time of programming of the gene as well as its subsequent expression (Rana and Gahan 1983a, Gahan and Carmignac 1987). Other gene switching can also be determined, together with changes of the levels of metabolic activity. This approach should help to determine the primary events in the differentiation of parenchyma cells in the presence or absence of initiating factors such as Ca^{2+} and growth regulators.

Fukuda and Komamine (1985) subdivided xylogenesis in the *Zinnia* system into four phases: I (0–24 h), II (24–48 h), III (48–72 h), and IV (72–96 h). In this brief epilog an examination of the earliest events may be instructive and point to some future areas of research. Determination is thought to require auxin plus a certain threshold concentration of cytokinin. For example, NAA (0.1 mg l^{-1}) plus BAP (1.0 µg l^{-1}) supports cell division but not xylogenesis, whereas NAA (0.1 mg l^{-1}) plus BAP (100 µg l^{-1}) induces both processes (Masuda et al. 1983). Auxin and cytokinin are known to synergize ethylene biosynthesis (Fuchs and Lieberman 1968, Lieberman 1979), and evidence that ethylene is a third hormone in determination has been outlined (Chap. 5.5, 5.6). One hypothesis that should be examined is that a threshold level of cytokinin is necessary in order to induce the biosynthesis of a xylogenic concentration of ethylene.

Does determination involve some sequential order of hormones? Some evidence indicates that the requirements for auxin persists longer than the requirement for cytokinin (Aloni 1982, Tucker et al. 1986). On the other hand, Phillips (1987) has taken a different viewpoint based on the induction of xylogenesis in Jerusalem artichoke explants given sequential treatments of 2,4-D and BAP. According to his interpretation, auxin was necessary early in the 4-day culture period for cellular proliferation. Cytokinin stimulated xylogenesis when present during the last 2 days of culture, that is, coincident with the expression of cytodifferentiation. Sequential hormone treatments involve inherent flaws, and it has been very difficult to resolve these problems. An auxin-alone treatment may involve endogenous cytokinin, and a "removal" type of experiment has carry-over effects (Phillips 1987).

One area that needs further investigation involves the roles of intracellular Ca^{2+} in xylogenesis. Since calmodulin-binding drugs were highly effective in completely blocking in vitro xylogenesis (Roberts and Baba 1987), presumably Ca^{2+} has a role in the initiation process as an auxin-regulated second messenger. Calcium may also be an important factor in the establishement of secondary wall patterns associated with developing tracheary elements. The application of a calmodulin-blocking agent during the latter phases of auxin-induced xylogenesis (72 h of culture) resulted in tracheary elements with highly irregular patterns of wall thickening (Roberts and Baba 1987). It is not surprising that the "patterning mechanism" is sensitive to the level of intracellular Ca^{2+}. There has been general support for the concept that cortical microtubules regulate the orientation of wall microfibrils, and elevated levels of Ca^{2+} are instrumental in microtubule disassembly (Gunning and Hardham 1982). Still another suggested role is the regulation of actin-myosin by Ca^{2+} (Williamson 1986). A rearrangement of actin filaments occurs in xylogenic *Zinnia* cells during the second phase (24–48 h), and this may be the mechanism for the shifting of microtubule arrays within the cell (Kobayashi et al. 1987). Reports involving organelle movements along actin filaments and microtubules has been reviewed (Williamson 1986).

There is a need for further research on the control of phloem differentiation and its role in the control of xylem differentiation. It should be recalled that, within the intact plant, xylem does not differentiate in the absence of phloem (Aloni 1987b). It is important to clarify whether or not there is a stimulus for xylem differentiation which might arrive through the phloem. Experimental evidence reveals that phloem differentiates with no xylem under limiting conditions, for example, incubation at 15 °C (Gautheret 1966) or in the presence of low auxin levels (Aloni 1980, Comer and Leonardo 1981). In addition, there is a requirement for higher auxin levels for the induction of xylem in comparison to the initiation of phloem formation. Aloni (1980) suggested that phloem is formed in response to auxin only, whereas xylem is formed in response to both auxin plus a stimulus which comes via the phloem. Experimental results obtained by Savidge and Wareing (1981a) led them to suggest that an unidentified tracheid differentiation factor (TDF) produced in coniferous needles limits xylem production. Results by Ewers and Aloni (1985) supported the concept of a hypothetical TDF. Further research is needed in order to establish the existence of such a factor. Liquid cultures of isolated cell suspensions may be helpful in solving this problem. If, in such a system, xylem elements can be induced in the absence of phloem elements, one may identify a specific factor needed for xylem differentiation in the absence of phloem. New methods will be required to identify any isolated sieve elements in the cultures. This can be done with the use of antibodies against specific proteins found in sieve elements (Sjolund 1987).

References

Akashi T, Shibaoka H 1987 Effects of gibberellin on the arrangement and cold stability of cortical microtubules in epidermal cells of pea internodes. Plant Cell Physiol 28:339–348

Albersheim P, Darville A G, McNeil M, Valent B, Sharp J K, Nothnagel E A, Davis K R, Yamazaki N, Gillin D J, York W S, Dudman W F, Darvill J E 1983 Oligosaccharins: Naturally occurring carbohydrates with biological regulatory functions. In: Ciferri O (ed) Structure and function of plant genomes. Plenum, New York, pp 293–312

Albinger G, Beiderbeck R 1983 Differentiation of tracheary elements and storage parenchyma cells in suspension cultures of Asparagus plumosus Bake. Z Pflanzenphysiol 112:443–448

Allen J R F 1980 Vascular transport of hormones in Ricinus communis L. Ph D thesis, Univ Sussex, England

Allen J R F, Baker D A 1980 Free tryptophan and indole-3-acetic acid levels in the leaves and vascular pathways of Ricinus communis L. Planta 148:69–74

Allen J R F, Greenway A M, Baker D A 1979 Determinations of indole-3-acetic acid in xylem sap of Ricinus communis L. using mass fragmentography. Planta 144:299–303

Aloni R 1976a The acropetal effect of indoleacetic acid on vascular differentiation and shoot development in Coleus blumei Benth. Isr J Bot 25:195–202

Aloni R 1976b Regeneration of phloem fibres round a wound: A new experimental system for studying the physiology of fibre differentiation. Ann Bot 40:395–397

Aloni R 1976c Polarity of induction and pattern of primary phloem fiber differentiation in Coleus. Am J Bot 63:877–889

Aloni R 1978 Source of induction and sites of primary phloem fibre differentiation in Coleus blumei. Ann Bot 42:1261–1269

Aloni R 1979 Role of auxin and gibberellin in differentiation of primary phloem fibers. Plant Physiol 63:609–614

Aloni R 1980 Role of auxin and sucrose in the differentiation of sieve and tracheary elements in plant tissue cultures. Planta 150:255–263

Aloni R 1982 Role of cytokinin in differentiation of secondary xylem fibers. Plant Physiol 70:1631–1633

Aloni R 1985 Plant growth method and composition. United States patent no. 4,507,144 issued March 26, 1985

Aloni R 1987a The induction of vascular tissues by auxin. In: Davies P J (ed) Plant hormones and their role in plant growth and development. Nijhoff/Junk, The Hague, pp 363–374

Aloni R 1987b Differentiation of vascular tissues. Ann Rev Plant Physiol 38:179–204

Aloni R, Gad A E 1982 Anatomy of the primary phloem fiber system in Pisum sativum. Am J Bot 69:979–984

Aloni R, Indig E F 1985 The control of vessel diameter in the stem of Luffa. Am J Bot 72:806 (abst)

Aloni R, Jacobs W P 1977a Polarity of tracheary regeneration in young internodes of Coleus (Labiatae). Am J Bot 64:395–403

Aloni R, Jacobs W P 1977b The time course of sieve tube and vessel regeneration and their relation to phloem anastomoses in mature internodes of Coleus. Am J Bot 64:615–621

Aloni R, Plotkin T 1985 Wound-induced and naturally occurring regenerative differentiation of xylem in Zea mays L. Planta 163:126–132

Aloni R, Raviv A, Wolf A 1986 Effect of light intensity on the differentiation of phloem anastomoses in cucumber seedlings. Am J Bot 73:618 (abst)

Aloni R, Sachs T 1973 The three-dimensional structure of primary phloem systems. Planta 113:345–353

Aloni R, Saks Y 1984 Differentiation of regenerative tracheids around a wound in Pinus. Am J Bot 71(Suppl):17 (Abstr)

Aloni R, Wolf A 1984 Suppressed buds embedded in the bark across the bole and the occurrence of their circular vessels in Ficus religiosa. Am J Bot 71:1060–1066

Aloni R, Zimmermann M H 1983 The control of vessel size and density along the plant axis – a new hypothesis. Differentiation 24:203–208

Aloni R, Zimmermann M H 1984 Length, width and pattern of regenerative vessels along strips of vascular tissue. Bot Gaz 145:50–54

Amer M, Neville P 1981 Morphogenesis in Gleditisia triacanthos L. Various growth substances effects on the percentage of xylem and phloem in the secondary tissue. C R Acad Sci 292:871–874

Amino S, Takeuchi Y, Komamine A 1985a Changes in intracellular UDP-sugar levels during the cell cycle in a synchronous culture of Catharanthus roseus. Physiol Plant 64:197–201

Amino S, Takeuchi Y, Komamine A 1985b Changes in synthetic activity of cell walls during the cell cycle in a synchronous culture of Catharanthus roseus. Physiol Plant 64:202–206

Ananiev E D, Karagyozov L K, Karanov E N 1987 Effect of cytokinins on ribosomal RNA gene expression in excised cotyledons of Cucurbita pepo L. Planta 170:370–378

Anderson L W J 1978 Abscisic acid induces formation of floating leaves in the heterophyllous aquatic angiosperm Potamogeton nodosus. Science 201:1135–1138

Artschwager E 1950 The time factor in the differentiation of secondary xylem and phloem in pecan. Am J Bot 37:15–24

Ashworth E N 1984 Xylem development in Prunus flower buds and the relationship to deep supercooling. Plant Physiol 74:862–865

Atal C K 1961 Effect of gibberellin on the fibers of hemp. Econ Bot 15:133–139

Avanzi S, Maggini F, Innocenti A M 1973 Amplification of ribosomal cistrons during the maturation of metaxylem in the roots of Allium cepa. Protoplasma 76:197–210

Baas P ed 1982 New perspectives in wood anatomy. Nijhoff/Junk, The Hague, p 252

Baas P, Carlquist S 1985 A comparison of the ecological wood anatomy of the floras of southern California and Israel. IAWA Bull N S 6:349–353

Baas P, Schmid R, Van Heuven B J 1986 Wood anatomy of Pinus longaeva (bristlecone pine) and the sustained length-on-age increase of its tracheids. IAWA Bull N S 7:221–228

Baas P, Werker E, Fahn A 1983 Some ecological trends in vessel characters. IAWA Bull N S 4:141–159

Baas P, Xinying Z 1986 Wood anatomy of trees and shrubs from China. 1. Oleaceae. IAWA Bull N S 7:195–220

Bachman B J 1982 Linkage map of Escherichia coli K-12. Genet Maps 2:95–107

Bacic A, Klein A S, Delmer D P 1980 Stimulation of glucan synthesis in vitro by induction of membrane potentials. Plant Physiol (Suppl):105 (Abstr)

Bailey I W 1958 The structure of tracheids in relation to the movement of liquids, suspensions and undissolved gases. In: Thimann K V (ed) The physiology of forest trees. Ronald, New York, pp 71–82

Bandurski R S, Nonhebel H M 1984 Auxins. In: Wilkins M B (ed) Advanced plant physiology. Pitman, London, pp 1–20

Banko T J, Roberts L W, Boe A A 1976 The detection of cytokinins in xylem exudate of Coleus blumei Benth. by the induction of xylogenesis in pith parenchyma explants of Lactuca sativa L. Ann Bot 40:651–654

Bannan M W 1950 The frequency of anticlinal divisions in fusiform cambial cells of Chamaecyparis. Am J Bot 37:511–519

Bannan M W 1955 The vascular cambium and radial growth in Thuja occidentalis L. Can J Bot 33:113–138

Barlow P W 1976 Towards an understanding of the behaviour of root meristems. J Theor Biol 57:433–451

Barlow P W 1977 The dynamic aspect of cytodifferentiation in cells. Naturwissenschaften 64:532

Barlow P W 1978 Endopolyploidy: Towards an understanding of its biological significance. Acta Theor 27:1–18

Barlow P W 1984 Positional controls in root development. In: Barlow P W, Carr D J (eds) Positional controls in plant development. Cambridge University Press, Cambridge, pp 281–318

Barnett J R 1981 Secondary xylem cell development. In: Barnett J R (ed) Xylem cell development. Castle House, Tunbridge Wells, pp 45–95

Barnett J R 1982 Plasmodesmata and pit development in secondary xylem elements. Planta 155:251–260

Barondes S H 1981 Lectins: Their multiple endogenous cellular functions. Ann Rev Plant Physiol 50:207–231

Bassi P K, Spencer M S 1982 Effect of carbon dioxide and light on ethylene production in intact sunflower plants. Plant Physiol 69:1222–1225

Bassi P K, Spencer M S 1983 Does light inhibit ethylene production in leaves? Plant Physiol 73:758–760

Baydoun E A H, Northcote D H 1980a Isolation and characterization of membranes from the cells of maize root tips. J Cell Sci 45:147–167

Baydoun E A H, Northcote D H 1980b Measurement and characteristics of fusion of isolated membrane fractions from maize root tips. J Cell Sci 45:169–186

Baydoun E A H, Northcote D H 1981 The extraction from maize (Zea mays) root cells of membrane-bound protein with Ca^{2+}-dependent ATPase activity and its possible role in membrane fusion in vitro. J Biochem 193:781–791

Beach D, Durkacz B, Nurse P 1982 Functionally homologous cell cycle control genes in budding and fission yeast. Nature 300:706–709

Becerra J, López-Sáez J F 1978 Effects of caffeine, calcium and magnesium on plant cytokinesis. Exp Cell Res 111:301–308

Behnke H D, Schulz A 1980 Fine structure, pattern of division, and course of wound phloem in Coleus blumei. Planta 150:357–365

Behnke H D, Sukkri B 1971 Anastomoses in the internode of Dioscorea: Their frequency, structure, and function. Z Pflanzenphysiol 66:82–92

Benayoun J, Aloni R, Sachs T 1975 Regeneration around wounds and the control of vascular differentiation. Ann Bot 39:447–454

Bennett M D 1981 The spatial order of barley chromosomes. Proc 4th Int Barley Symp, Edinburgh, pp 751–757

Bennett M D 1982 Nucleotype basis of the spatial ordering of chromosomes in eukaryotes and the implications of the order for genome evolution and phenotypic variation. In: Dover G A, Flavell R B (eds) Genome evolution. Academic Press, New York, London, pp 239–261

Bennett M D 1983 The spatial distribution of chromosomes. In: Brandham P, Bennett M D (eds) Proc Kew chromosome conference II. Allen and Unwin, London, pp 71–79

Bennett M D, Smith J B 1976 Nuclear DNA amounts in angiosperms. Philos Trans R Soc London Ser B 274:227–274

Benochea T, Harry G I, Dodds J H 1983 Cytodifferentiation of xylem cells in vitro – an overview. Histochem J 15:411–418

Benson M 1910 Root parasitism in Exocarpus (with comparative notes on the haustoria of Thesium). Ann Bot 24:667–677

Bergmann L 1964 Der Einfluß von Kinetin auf die Ligninbildung und Differenzierung in Gewebekulturen von Nicotiana tabacum. Planta 62:221–254

Berlyn G P 1979 Physiological control of differentiation of xylem elements. Wood Fiber 11:109–126

Berlyn G P 1982 Morphogenetic factors in wood formation and differentiation. In: Baas P (ed) New perspectives in wood anatomy. Nijhoff/Junk, The Hague, pp 123–150

Berlyn G P, Battey Y C 1985 Metabolism and synthetic function of cambial tissue. In: Higuchi T (ed) Biosynthesis and biodegradation of wood components. Academic Press, New York, London, pp 63–85

Berridge M J 1976 Calcium, cyclic nucleotides and cell division. In: Duncan C J (ed) Calcium in biological systems. Cambridge University Press, Cambridge, pp 219–231

Bevan M, Northcote D H 1979a The loss of morphogenetic potential and induction of phenylalanine ammonia-lyase in suspension cultures of Phaseolus vulgaris. J Cell Sci 39:339–353

Bevan M, Northcote D H 1979b The interaction of auxin and cytokinin in the induction of phenylalanine ammonia-lyase in suspension cultures of Phaseolus vulgaris. Planta 147:77–81

Bevan M, Northcote D H 1981a Subculture-induced protein synthesis in tissue cultures of Glycine max and Phaseolus vulgaris. Planta 152:24–31

Bevan M, Northcote D H 1981 b Some rapid effects of synthetic auxins on RNA levels in cultured plant cells. Planta 152:32–35

Beyer E M, Jr 1979 Effect of silver ion, carbon dioxide, and oxygen on ethylene action and metabolism. Plant Physiol 63:169–173

Binns A N, Chen R H, Wood H N, Lynn D G 1987 Cell division promoting activity of naturally occurring dehydrodiconiferyl glucosides: Do cell wall components control cell division? Proc Natl Acad Sci USA 84:980–984

Biro R L, Hunt E R, Erner Y, Jaffe M J 1980 Thigmomorphogenesis: Changes in cell division and elongation in the internodes of mechanically-perturbed or ethrel-treated bean plants. Ann Bot 45:655–664

Biro R L, Jaffe M J 1984 Thigmomorphogenesis – ethylene evolution and its role in the changes observed in mechanically perturbed bean plants. Physiol Plant 62:289–296

Bisalputra T, Esau K 1964 Polarized light study of phloem differentiation in embryo of Chenopodium album. Bot Gaz 125:1–7

Blackmore S, Tootill E eds 1984 The Penguin dictionary of botany. Penguin Books, Harmondsworth, Middlesex, 391 pp

Blakely L M, Blakely R M, Galloway C M 1986 Effects of dimethyl sulfoxide and pH on indoleacetic acid-induced lateral root formation in the radish seedling root. Plant Physiol 80:790–791

Bolwell G P 1985 Use of tissue cultures for studies on vascular differentiation. In: Dixon R A (ed) Plant cell culture. A practical approach. IRL, Oxford, pp 107–126

Bolwell G P, Dalessandro G, Northcote D H 1985 Decrease of polygalacturonic acid synthase during xylem differentiation in sycamore. Phytochemistry 24:699–702

Bolwell G P, Northcote D H 1981 Control of hemicellulose and pectin synthesis during differentiation of vascular tissue in bean (Phaseolus vulgaris) callus and in bean hypocotyl. Planta 152:225–233

Bolwell G P, Northcote D H 1983 a Arabinan synthase and xylan synthase activities of Phaseolus vulgaris. Subcellular localization and possible mechanism of action. Biochem J 210:497–507

Bolwell G P, Northcote D H 1983 b Induction by growth factors of polysaccharide synthases in bean cell suspension cultures. Biochem J 210:509–515

Bonnemain J L 1971 Transport et distribution des traceurs après application de AIA-2-^{14}C sur les feuilles de Vicia faba. C R Acad Sci 273D:1699–1702

Bornman C H 1983 Morphological, anatomical, and ultrastructural responses to abscisic acid. In: Addicott F T (ed) Abscisic acid. Praeger, New York, pp 523–551

Bornman C H 1985 Hormonal control of growth and differentiation in conifer tissues in vitro. Biol Plant 27:249–256

Bornman C H, Huber W 1979 Nicotiana tabacum callus studies. IX. Development in stressed explants. Biochem Physiol Pflanz 174:345–356

Bower F O 1930 Size and form in plants. Macmillan, London, 232 pp

Bower F O 1935 Primitive land plants. Macmillan, London, 658 pp

Boyer N, Chapelle B, Gaspar T 1979 a Lithium inhibition of the thigmomorphogenetic response in Bryonia dioica. Plant Physiol 63:1215–1216

Boyer N, De Jaegher G, Bon M C, Gaspar T 1986 Cobalt inhibition of thigmomorphogenesis in Bryonia dioica. Possible role and mechanism of ethylene production. Physiol Plant 67:552–556

Boyer N, Desbiez M O, Hofinger M, Gaspar T 1983 Effect of lithium on thigmomorphogenesis in Bryonia dioica. Ethylene production and sensitivity. Plant Physiol 72:522–525

Boyer N, Gaspar T, Lamond M 1979 b Changes in isoperoxidases and in the growth rate of Bryonia dioica internodes after mechanical rubbing. Z Pflanzenphysiol 93:459–470

Brabec F 1953 Über Polysomatie in der Wurzel von Bryonia. Chromosoma 6:135–141

Bradford K J, Yang S F 1980 Xylem transport of 1-amino-cyclopropane-1-carboxylic acid, an ethylene precursor, in waterlogged tomato plants. Plant Physiol 65:322–326

Bradley M V, Dahmen W J 1971 Cytohistological effects of ethylene, 2,4-D, kinetin and carbon dioxide on peach mesocarp callus cultured in vitro. Phytomorphology 21:154–164

Braton B O, Henry E W 1977 Electrical stimulation and its effects on indoleacetic acid and peroxidase levels in tomato plants (Lycopersicon esculentum). J Exp Bot 28:338–344

Brown C L 1964 The influence of external pressure on the differentiation of cells and tissues cultured in vitro. In: Zimmermann M H (ed) The formation of wood in forest trees. Academic Press, New York, London, pp 389–404

Brown C L, Sax K 1962 The influence of pressure on the differentiation of secondary tissues. Am J Bot 49:683–691

Brown E G, Newton R P 1981 Cyclic AMP and higher plants. Phytochemistry 20:2453–2463

Brown K M, Leopold A C 1973 Ethylene and the regulation of growth in pine. Can J For Res 3:142–145

Brown R 1976 Significance of division in the higher plants. In: Yeoman M M (ed) Cell division in higher plants. Academic Press New York, London, pp 3–46

Bruck D K, Paolillo D J, Jr 1984a Anatomy of nodes vs. internodes in Coleus: The nodal cambium. Am J Bot 71:142–150

Bruck D K, Paolillo D J, Jr 1984b Anatomy of nodes vs. internodes in Coleus: The longitudinal course of xylem differentiation. Am J Bot 71:151–157

Bruck D K, Paolillo D J, Jr 1984c The control of vascular branching in Coleus. 1. The side bundle. Ann Bot 53:727–736

Bruck D K, Paolillo D J, Jr 1984d The control of vascular branching in Coleus. 2. The corner traces. Ann Bot 53:737–747

Bruck D K, Paolillo D J, Jr 1984e Replacement of leaf primordia with IAA in the induction of vascular differentiation in the stem of Coleus. New Phytol 96:353–370

Brunori A 1971 Synthesis of DNA and mitosis in relation to cell differentiation in the roots of Vicia faba and Lactuca sativa. Caryologia 24:209–215

Bryans C, Smith D L 1985 Endopolyploidy, cell volume and nuclear volume interrelationships in cotyledons of the Leguminosae. Ann Bot 56:225–238

Bryant J A 1976 Molecular aspects of gene expression in plants. Academic Press, New York, London, 338 pp

Bryant J A, Francis D eds 1985 The cell division cycle in plants. Cambridge University Press, Cambridge, 258 pp

Bryne J M 1973 The root apex of Malva sylvestris. III. Lateral root development and the quiescent center. Am J Bot 60:657–662

Bucknall R A, Sutcliffe J F 1965 The nucleic acids and cellular differentiation in the root apex of Pisum sativum. II. The nucleic acids and cellular development patterns. J Exp Bot 16:587–603

Buckout T J, Young K A, Low P S, Morré D J 1981 In vitro promotion by auxins of divalent ion release from soybean membranes. Plant Physiol 68:512–515

Burgess J, Linstead P J 1982 Cell-wall differentiation during growth of electrically polarized protoplasts of Physcomitrella. Planta 156:241–248

Burgess J, Linstead P 1984 In vitro tracheary element formation: Structural studies and the effect of tri-iodobenzoic acid. Planta 160:418–489

Burke M J, Stushnoff C 1979 Frost hardiness: A discussion of the possible molecular causes of injury with particular reference to deep supercooling of water. In: Mussel H, Staples R C (eds) Stress physiology in crop plants. John Wiley, New York, pp 199–226

Butterfield B G, Meylan B A 1982 Cell wall hydrolysis in the tracheary elements of the secondary xylem. In: Baas P (ed) New perspectives in wood anatomy. Nijhoff/Junk, The Hague, pp 71–84

Caligo M A, Nuti Ronchi V, Nozzolini M 1986 Proline and serine affect polarity and development of carrot somatic embryos. Cell Differ 17:193–198

Caplan A I, Ordall C P 1978 Irreversible gene repression model for control of development. Science 201:120–130

Carlquist S 1977 Ecological factors in wood evolution: a floristic approach. Am J Bot 64:887–896

Carlquist S, Hoekman D A 1985 Ecological wood anatomy of the woody southern California flora. IAWA Bull N S 6:319–347

Carmignac D M, Gahan P B 1987 Switching of carboxylesterase activities during the establishment of the procambium and procortex in pea embryos (submitted)

Catesson A M 1964 Origine, fonctionnement et variations cytologiques saisonnières du cambium de l'Acer pseudoplatanus L. (Acéracées). Ann Sci Nat Bot 12 Ser 5:229–298

Catesson A M 1974 Cambial cells. In: Robards A W (ed) Dynamic aspects of plant ultrastructure. McGraw-Hill, London, pp 358–390

Catesson A M 1980 The vascular cambium. In: Little C H A (ed) Control of shoot growth in trees. Proc IUFRO Workshop, Fredericton, Canada, pp 12–40

Catesson A M 1984 La dynamique cambiale. Ann Sci Nat Bot Biol Veg 6:23–43

Catesson A M, Czaninski Y, Monties B 1978 Charactères histochimiques des peroxidases pariétales dans les cellules en cours de lignification. C R Acad Sci 286D:1787–1790

Cawthon G E 1972 The role of cytokinin and factors affecting xylem differentiation in Romaine lettuce pith cultured in vitro. Ph D thesis, Univ Idaho

Chadwick C M, Garrod D R eds 1986 Hormones, receptors, and cellular interactions in plants. Cambridge University Press, Cambridge, 375

Chafe S C, Durzan D J 1973 Tannin inclusions in cell suspension cultures of white spruce. Planta 113:251–262

Chaleff R S 1981 Genetics of higher plants. Applications of cell culture. Cambridge University Press, Cambridge, 176 pp

Chalutz E 1973 Ethylene-induced phenylalanine ammonia-lyase activity in carrot roots. Plant Physiol 51:1033–1036

Chandra S V, Johri A K, Krishnan P N, Vaishnov P P, Singh Y D 1986 Distribution of indoleacetic acid oxidase, peroxidase and esterase activities in relation to elongation in pearl millet internode. J Plant Physiol 122:373–383

Chapman G P, Mantell S H, Daniels R W 1985 Experimental manipulation of ovule tissues. Longman, London, 256 pp

Chapman J M, Perry R 1987 The diffusion theory of phyllotaxis. Ann Bot 60:377–389

Chauhan Y S, Roberts L W 1978 Tracheary elements in vesicles of Citrus fruits from northeastern India. Phytomorphology 28:96–98

Chriqui D, Dorazi D, Bagni N 1986 Ornithine and arginine decarboxylase and polyamine involvement during in vivo differentiation and in vitro dedifferentiation of Datura innoxia leaf explants. Physiol Plant 68:589–596

Christoffersen R E, Laties G G 1982 Ethylene regulation of gene expression in carrots. Proc Natl Acad Sci USA 79:4060–4063

Christoffersen R E, Warm E, Laties G G 1982 Gene expression during fruit ripening in avocado. Planta 155:52–57

Clowes F A L 1958 Development of quiescent centres in root meristems. New Phytol 57:85–88

Clowes F A L 1961 Apical meristems. Blackwell, Oxford, 217 pp

Clowes F A L 1971 The proportion of cells that divide in root meristems of Zea mays L. Ann Bot 35:249–261

Clowes F A L 1972 The control of cell proliferation within root meristems. In: Miller M W, Kuehnert CC (eds) The dynamics of meristem cell populations. Plenum, New York, pp 133–147

Clowes F A L 1978 Origin of the quiescent centre in Zea mays. New Phytol 80:409–419

Clutter M E 1960 Hormonal induction of vascular tissue in tobacco pith in vitro. Science 132:548–549

Cohen J D, Nadler K D 1976 Calcium reqirement for IAA-induced acidification by Avena coleoptiles. Plant Physiol 57:347–350

Cohen J D, Meudt W J 1983 Investigation on the mechanism of the brassinosteroid response. I. Indole-3-acetic acid metabolism and transport. Plant Physiol 72:691–694

Cohn N S, Mitchell J P, van den Broek K 1983 Nuclear and nucleolar protein during the cell cycle in differentiating Pisum sativum vascular tissue. Histochemistry 79:353–364

Comer A E 1978 Pattern of cell division and wound vessel member differentiation in Coleus pith explants. Plant Physiol 62:354–359

Comer A E, Leonardo L 1981 Phloem differentiation in Coleus pith explants. Physiol Plant 51:130–132

Comings D E 1968 The rationale for an ordered arrangement of chromatin in the interphase nucleus. Am J Hum Genet 20:440–460

Comings D E 1980 Arrangement of chromatin in the nucleus. Hum Genet 53:131–143

Cookson C, Osborne D J 1978 The stimulation of cell extension by ethylene and auxin in aquatic plants. Planta 144:39–47

Cooper S 1979 A unifying model for a G_1 period in prokaryotes and eukaryotes. Nature 280:17–19

Cooper S 1981 The central dogma of cell biology. Cell Biol Int Rep 5:539–549

Corsi G, Avanzi S 1970 Cytochemical analyses of cellular differentiation in the root tip of Allium cepa. Caryologia 23:381–394

Craker L E, Wetherbee P J 1973 Ethylene, carbon dioxide, and anthocyanin synthesis. Plant Physiol 52:177–179

Creber G T, Chaloner W G 1984 Influence of environmental factors on the wood structure of living and fossil trees. Bot Rev 50:357–448

Cross J W 1985 Auxin action: the search for the receptor. Plant, Cell and Environment 8:351–359

Cumbie B G 1967 Development and structure of the cambium of Canavalia. Bull Torrey Bot Club 94:162–175

Cutter E G 1971 Plant anatomy: Part 2 Organs. Experiment and interpretation. Edward Arnold, London, 343 pp

Cutter E G 1978 Plant anatomy: Part 1 Cells and tissues. Experiment and interpretation. Edward Arnold, London, 168 pp

Dahl G, Schuldt C, Gratzl M 1978 Fusion of isolated myoblast plasma membranes. An approach to the mechanism. Biochim Biophys Acta 514:105–116

Dalessandro G 1973 Interaction of auxin, cytokinin, and gibberellin on cell division and xylem differentiation in cultured explants of Jerusalem artichoke. Plant Cell Physiol 14:1167–1176

Dalessandro G, Northcote D H 1977a Changes in enzymic activities of nucleoside diphosphate sugar interconversions during differentiation of cambium to xylem in sycamore and poplar. Biochem J 162:267–279

Dalessandro G, Northcote D H 1977b Changes in enzymic activities of nucleoside diphosphate sugar interconversions during differentiation of cambium to xylem in pine and fir. Biochem J 162:281–288

Dalessandro G, Northcote D H 1977c Changes in enzymic activities of UDP-D-glucuronate decarboxylase and UDP-D-xylose 4-epimerase during cell division and xylem differentiation in cultured explants of Jerusalem artichoke. Phytochemistry 16:853–859

Dalessandro G, Northcote D H 1981a Xylan synthetase activity in differentiated xylem cells of sycamore trees (Acer pseudoplatanus). Planta 151:53–60

Dalessandro G, Northcote D H 1981b Increase of xylan synthetase activity during xylem differentiation of the vascular cambium of sycamore and poplar trees. Planta 151:61–67

Dalessandro G, Piro G, Northcote D H 1986 Glucomannan-synthase activity in differentiating cells of Pinus sylvestris L. Planta 169:564–574

Dalessandro G, Roberts L W 1971 Induction of xylogenesis in pith parenchyma explants of Lactuca. Am J Bot 58:378–385

Datta S K, Chakrabarti K, Datta P C 1975 In vitro effect of acidity levels on xylem differentiation in Plumeria. Curr Sci 44:814–816

Davies W J 1987 Hormones as chemical signals involved in root to shoot communication of changes in the soil environment. XIV Int Bot Congr, Berlin (Abstr 6-30-4)

Davis D E 1949 Some effects of calcium deficiency on the anatomy of Pinus taeda. Am J Bot 36:276–282

Dawson A L, Gahan P B 1979 Longevity of hydrolase molecules during primary xylem differentiation. Ann Bot 43:251–254

Dedman J R, Brinkley B R, Means A R 1979 Regulation of microfilaments and microtubules by calcium and cyclic AMP. Adv Cycl Nucl Res 11:131–174

Deeley E M, Davies H G, Chayen J 1957 The DNA content of cells in the root of V. faba. Exp Cell Res 12:582–591

De Guzman C C, Dela Fuente R K 1984 Polar calcium flux in sunflower hypocotyl segments. I. The effect of auxin. Plant Physiol 76:347–352

De Jaegher G, Boyer N 1987 Specific inhibition of lignification in Bryonia dioica. Effects on thigmomorphogenesis. Plant Physiol 84:10–11

De Jaegher G, Boyer N, Bon M C, Gaspar T 1987 Thigmomorphogenesis in Bryonia dioica: early events in ethylene biosynthesis pathway. Biochem Physiol Pflanz 182:49–56

De Jaegher G, Boyer N, Gaspar T 1985 Thigmomorphogenesis in Bryonia dioica: changes in soluble and wall peroxidases, phenylalanine ammonia-lyase activity, cellulose, lignin content and monomeric constituents. Plant Growth Regul 3:133–148

Dela Fuente R K 1984 Role of calcium in the polar secretion of indoleacetic acid. Plant Physiol 76:342–346

Delmer D P, Thelen M, Marsden M P F 1984 Regulatory mechanism for the synthesis of β-glucans in plants. In: Dugger W M, Bartnicki-Garcia S (eds) Structure, function, and biosynthesis of plant cell walls. Am Soc Plant Physiol, Rockville, pp 133–149

DeMaggio A E 1966 Phloem differentiation: induced stimulation by gibberellic acid. Science 152:370–372

Denne M P 1971 Temperature and tracheid development in Pinus sylvestris seedlings. J Exp Bot 22:362–370

Denne M P 1972 A comparison of root and shoot wood development in conifer seedlings. Ann Bot 36:579–587

Denne M P 1977 Some effects of wounding on tracheid differentiation in Picea sitchensis. IAWA Bull 3:49–50

Denne M P, Dodd R S 1981 The environmental control of xylem differentiation. In: Barnett J R (ed) Xylem cell development. Castle House, Tunbridge Wells, pp 236–255

Denne M P, Whitbread V 1978 Variation of fibre length within trees of Fraxinus excelsior. Can J For Res 8:253–260

Denne M P, Wilson J E 1977 Some quantitative effects of indoleacetic acid on the wood production and tracheid dimensions of Picea. Planta 134:223–228

Dieter P, Marmé D 1980 Calmodulin-activated plant microsomal Ca^{2+} uptake and purification of plant NAD kinase and other proteins by calmodulin-sepharose chromatography. In: Siegel F L, Carafoli E, Kretsinger R H, MacLennan D H, Wasserman R H (eds) Calcium binding proteins: structure and function. Elsevier/North Holland, Amsterdam, New York, pp 211–212

Digby J, Wareing P F 1966 The effect of applied growth hormones on cambial division and the differentiation of the cambial derivatives. Ann Bot 30:539–548

Dini F, Viola-Magni M P 1976 DNA metabolic instability during hydrolysis. Biochim Biol Sper 12:63–70

Dinwoodie J M 1961 Tracheid and fibre length in timber: A review of literature. Forestry 34:125–144

Dobbins D R, Fisher J B 1986 Wound responses in girdled stems of lianas. Bot Gaz 147:278–289

Dobbins D R, Kuijt J 1973 Studies on the haustorium of Castilleja (Scrophulariaceae). II The endophyte. Can J Bot 51:923–931

Dodd R S 1985 Within-tree variation in wood production and wood quality in sycamore (Acer pseudoplatanus): its relation to crown characteristics. Can J For Res 15:56–65

Dodds J H 1980 The effect of 5-fluorodeoxyuridine and colchicine on tracheary element differentiation in isolated mesophyll cells of Zinnia elegans L. Z Pflanzenphysiol 99:283–285

Dodds J H 1981 a The role of the cell cycle in differentiation, a new model. Plant Cell Environ 5:101–107

Dodds J H 1981 b The role of the cell cycle and cell division in xylem differentiation. In: Barnett J R (ed) Xylem cell development. Castle House, Tunbridge Wells, pp 153–167

Dodds J H, Phillips R 1977 DNA and histone content of immature tracheary elements from cultured artichoke explants. Planta 135:213–216

Dodds J H, Roberts L W 1985 Experiments in plant tissue culture, 2nd edn. Cambridge University Press, Cambridge, 232 pp

Doley D, Leyton L 1968 Effects of growth regulating substances and water potential on the development of secondary xylem in Fraxinus. New Phytol 67:579–594

Doley D, Leyton L 1970 Effects of growth regulating substances and water potential on the development of wound callus in Fraxinus. New Phytol 69:87–102

Dudley K, Northcote D H 1979 Regulation of induction of phenylalanine ammonia-lyase in suspension cultures of Phaseolus vulgaris. Planta 146:433–440

Dugger W M 1983 Boron in plant metabolism. In: Lauchli A, Bieleski R L (eds) Inorganic plant nutrition. Encycl Plant Physiol, New Ser, vol 15B. Springer, Berlin Heidelberg New York, pp 626–650

Durante M, Cremonini R, Brunori A, Avanzi S, Innocenti A M 1977 Differentiation of metaxylem cell line in roots of Allium cepa. 1. DNA heterogeneity and ribosomal cistrons of two different stages of differentiation. Protoplasma 93:289–303

Dure L 1985 Embryogenesis and gene expression during seed formation. Oxford Surv Plant Mol Cell Biol 2:179–197

Durzan D J, Chafe S C, Lopushanski S M 1973 Effects of environmental changes on sugars, tannins, and organized growth in cell suspension cultures of white spruce. Planta 133:241–249

Eames A J, MacDaniels L H 1947 An introduction to plant anatomy. McGraw-Hill, New York, 427 pp

Earle E D 1968 Induction of xylem elements in isolated Coleus pith. Am J Bot 55:302–305

Edmunds L N 1984 Cell cycle clocks. Marcell Dekker New York, 616 pp

Einset J W 1978 Citrus tissue culture. Stimulation of fruit explant cultures with orange juice. Plant Physiol 62:885–888

Elliott D C 1979a Differentiation effects on fusicoccin- and cytokinin-stimulated betacyanin synthesis by ATPase inhibitors and uncouplers. Plant Sci Lett 15:251–264

Elliott D C 1979b Temperature-sensitive responses of red light-dependent betacyanin synthesis. Plant Physiol 64:521–524

Elliott D C 1983 Inhibition of cytokinin-regulated responses by calmodulin-binding compounds. Plant Physiol 72:251–258

Elliott D C, Batchelor S M, Cassar R A, Marinos N G 1983 Calmodulin-binding drugs affect responses to cytokinin, auxin, and gibberellic acid. Plant Physiol 72:219–224

Erner Y, Jaffe M J 1982 Thigmomorphogenesis: the involvement of auxin and abscisic acid in growth retardation due to mechanical perturbation. Plant Cell Physiol 23:935–941

Erner Y, Reuveni O, Goldschmidt E E 1975 Partial purification of a growth factor from orange juice which affects citrus tissue culture and its replacement by citric acid. Plant Physiol 56:279–282

Esau K 1942 Vascular differentiation in the vegetative shoots of Linum. 1. The procambium. Am J Bot 29:738–747

Esau K 1943 Origin and development of primary vascular tissues in seed plants. Bot Rev 9:125–206

Esau K 1965a Plant anatomy. Second edition. John Wiley, New York, 767 pp

Esau K 1965b Vascular differentiation in plants. Holt Rinehart Winston, New York, 160 pp

Esau K 1969 The phloem. In: Zimmermann W, Ozenda P, Wulff HD (eds) Encyclopedia of plant anatomy, vol V, part 2. Borntraeger, Berlin, 505 pp

Esau K 1977 Anatomy of seed plants. Second edition. John Wiley, New York, 550 pp

Eschrich W 1953 Beiträge zur Kenntnis der Wundsiebröhrenentwicklung bei Impatiens holstii. Planta 43:37–74

Evans L S, Tramontano W A 1981 Is trigonelline a plant hormone? Am J Bot 68:1282–1289

Evans M L 1984 Functions of hormones at the cellular level of organization. In: Scott T K (ed) Hormonal regulation of development II. The functions of hormones from the level of the cell to the whole plant. Encycl Plant Physiol, New Ser, vol 10. Springer, Berlin Heidelberg New York, pp 23–79

Evert R F 1963 The cambium and seasonal development of the phloem in Pyrus malus. Am J Bot 50:149–159

Ewers F W 1982a Developmental and cytological evidence for mode of origin of secondary phloem in needle leaves of Pinus longaeva (bristlecone pine) and P. flexilis. Bot Jahrb Syst 103:59–88

Ewers F W 1982b Secondary growth in needle leaves of Pinus longaeva (bristlecon pine) and other conifers: quantitative data. Am J Bot 69:1552–1559

Ewers F W, Aloni R 1984 The hormonal control of secondary growth in leaves of Pinus. Am J Bot 71(Suppl):27 (Abstr)

Ewers F W, Aloni R 1985 The effects of applied auxin and gibberellin on phloem and xylem production in needle leaves of Pinus. Bot Gaz 146:466–471

Ewers F W, Aloni R 1987 Seasonal secondary growth in needle leaves of Pinus strobus and Pinus brutia. Am J Bot 74:980–987

Fahn A 1982 Plant anatomy, 3rd edn. Pergamon, Oxford, 544 pp

Fahn A, Ben-Sasson R, Sachs T 1972 The relation between the procambium and the cambium. In: Ghouse A K M, Yunus M (eds) Research trends in plant anatomy. Tata McGraw-Hill, Bombay, New Dehli, pp 161–174

Fahn A, Werker E, Baas P 1986 Wood anatomy and identification of trees and shrubs from Israel and adjacent regions. Isr Acad Sci, Jerusalem, 221 pp

Falconer M M, Seagull R W 1985a Immunofluorescent and Calcofluor white staining of developing tracheary elements in Zinnia elegans L. suspension cultures. Protoplasma 125:190–198

Falconer M M, Seagull R W 1985b Xylogenesis in tissue culture: Taxol effects on microtubule reorientation and lateral association in differentiating cells. Protoplasma 128:157–166

Falconer M M, Seagull R W 1986 Xylogenesis in tissue culture II. Microtubules, cell shape and secondary wall patterns. Protoplasma 133:140–148

Falconer M M, Seagull R W 1987 Amiprophos-methyl (APM): a rapid reversible, anti-microtubule agent for plant cell cultures. Protoplasma 136:118–124

Farzaneh F, Zalin R, Brill D, Shall S 1982 DNA strand breaks and ADP-ribosyl transferase activation during cell differentiation. Nature 300:362–366

Fensom D S, Meylan S, Pilet P E 1965 Induced electro-osmosis in root tissues. Can J Bot 43:453–467

Fielding J L, Hall J L 1978a A biochemical and cytochemical study of peroxidase activity in roots of Pisum sativum. I. A comparison of DAB-peroxidase and guaiacol-peroxidase with particular emphasis on the properties of cell wall activity. J Exp Bot 29:969–981

Fielding J L, Hall J L 1978b A biochemical and cytochemical study of peroxidase activity in roots of Pisum sativum. II. Distribution of enzymes in relation to root development. J Exp Bot 29:983–991

Fineran B A 1983 Ultrastructure of graniferous tracheary elements in the terrestrial mistletoe Nuytsia floribunda (Loranthaceae). Protoplasma 116:57–64

Fineran B A 1985 Graniferous tracheary elements in haustoria of root parasitic angiosperms. Bot Rev 51:389–441

Fineran B A, Ingerfeld M 1982 Graniferous tracheary elements in the haustorium of Atkinsonia ligustrina, a root hemi-parasite of the Loranthaceae. Protoplasma 113:150–160

Fineran B A, Ingerfeld M, Patterson W D 1987 Inclusions of graniferous tracheary elements in the root hemi-parasite Olax phyllanthi (Olacaceae). Protoplasma 136:16–28

Fineran B A, Juniper B E, Bullock S 1978 Graniferous tracheary elements in the haustorium of the Santalaceae. Planta 141:29–32

Fisher J B 1981 Wound healing by exposed secondary xylem in Adamsonia (Bombacaceae). IAWA Bull N S 2:193–199

Fisher J B 1985 Induction of reaction wood in Terminalia (Combretaceae): roles of gravity and stress. Ann Bot 55:237–248

Ford B J 1981 The annotation of Leeuwenhoeck's packets of specimens: a preliminary account. Proc R Microsc Soc 16:393–395

Fosket D E 1968 Cell division and the differentiation of wound vessel members in cultured stem segments of Coleus. Proc Natl Acad Sci USA 59:1089–1096

Fosket D E 1970 The time course of xylem differentiation and its relation to deoxyribonucleic acid synthesis in cultured Coleus stem segments. Plant Physiol 46:64–68

Fosket D E 1972 Meristematic activity in relation to wound xylem differentiation. In: Miller M W, Kuehnert C C (eds) The dynamics of meristem cell populations. Plenum, New York, pp 33–50

Fosket D E 1977 The regulation of the plant cell cycle by cytokinin. In: Rost T L, Gifford E M, Jr (eds) Mechanisms and control of cell division. Dowden Hutchinson and Ross, Stroudsville, pp 62–91

Fosket D E 1980 Hormonal control of morphogenesis in cultured tissues. In: Skoog F (ed) Plant growth substances 1979. Springer, Berlin Heidelberg New York, pp 362–369

Fosket D E, Morejohn L C, Westerling K E 1981 Control of growth by cytokinin: An examination of tubulin synthesis during cytokinin induced growth in cultured cells of Paul's scarlet rose. In: Guern J, Péaud-Lenoël C (eds) Metabolism and molecular activities of cytokinins. Springer, Berlin Heidelberg New York, pp 193–211

Fosket D E, Roberts L W 1964 Induction of wound-xylem differentiation in isolated Coleus stem segments in vitro. Am J Bot 51:19–25

Fosket D E, Tepfer D A 1978 Hormonal regulation of growth in cultured plant cells. In Vitro 14:61–75

Fosket D E, Torrey J G 1969 Hormonal control of cell proliferation and xylem differentiation in cultured tissues of Glycine max var Biloxi. Plant Physiol 44:871–880

Foster A S, Gifford E M, Jr 1974 Comparative morphology of vascular plants. 2nd edn. Freeman, San Francisco, 751 pp

Fowke L C 1987 Investigations of cell structure using cultured cells and protoplasts. In: Green C E, Somers D A, Hackett W P, Biesboer D D (eds) Plant tissue and cell culture. Alan R Liss, New York, pp 17–31

Fry S C 1985 Primary cell wall metabolism. In: Miflin B (ed) Oxford surveys of plant molecular and cell biology, vol 2. Oxford University Press, Oxford, pp 1–42

Fuchs Y, Lieberman M 1968 Effects of kinetin, IAA and gibberellin on ethylene production and their interaction in growth of seedlings. Plant Physiol 43:2029–2036

Fuhrer J, Kaur-Sawhney R, Shih L M, Galston A W 1982 Effects of exogenous 1,3-diaminopropane and spermidine on senescence of oat leaves. II. Plant Physiol 70:1597–1600

Fukuda H 1987 A change of tubulin synthesis in the process of tracheary element differentiation and cell division of isolated Zinnia mesophyll cells. Plant Cell Physiol 28:517–528

Fukuda H, Iwata N 1986 Establishment of a sensitized immunoblotting method for measuring plant tubulin content. Plant Cell Physiol 27:273–283

Fukuda H, Komamine A 1980a Direct evidence for cytodifferentiation to tracheary elements without intervening mitosis in a culture of single cells isolated from the mesophyll of Zinnia elegans. Plant Physiol 65:61–64

Fukuda H, Komamine A 1980b Establishment of an experimental system for the study of tracheary element differentiation from single cells isolated from the mesophyll of Zinnia elegans. Plant Physiol 65:57–60

Fukuda H, Komamine A 1981a Relationship between tracheary element differentiation and the cell cycle in single cells isolated from the mesophyll of Zinnia elegans. Physiol Plant 52:423–430

Fukuda H, Komamine A 1981b Relationship between tracheary element differentiation and DNA synthesis in single cells isolated from the mesophyll of Zinnia elegans. Analysis by inhibitors of DNA synthesis. Plant Cell Physiol 22:41–49

Fukuda H, Komamine A 1982a Lignin synthesis and its related enzymes as markers of tracheary element differentiation in single cells isolated from mesophyll of Zinnia elegans. Planta 155:423–430

Fukuda H, Komamine A 1982b Relationship between cytodifferentiation and the cell cycle. In: Fujiwara A (ed) Plant tissue culture 1982. Jap Assoc Plant Tissue Cult, Tokyo, pp 97–98

Fukuda H, Komamine A 1983 Changes in the synthesis of RNA and protein during tracheary element differentiation in single cells isolated from the mesophyll of Zinnia elegans. Plant Cell Physiol 24:603–614

Fukuda H, Komamine A 1985 Cytodifferentiation. In: Vasil I K (ed) Cell culture and somatic cell genetics of plants, vol 2. Academic Press, New York, London, pp 149–212

Gad A E, Aloni R 1984 Time course and polarity of primary phloem fibre differentiation in Pisum sativum. Ann Bot 54:423–428

Gahan P B 1976 DNA turnover in nuclei of cells from higher plants. Riv Istochim Norm Patol 20:108–109

Gahan P B 1981 An early cytochemical marker of commitment to stelar differentiation in meristems from dicotyledonous plants. Ann Bot 48:769–775

Gahan P B 1984 Plant histochemistry and cytochemistry. An introduction. Academic Press, London, 301 pp

Gahan P B, Bellani L M 1984 Identification of shoot apical meristem cells to form vascular elements in Pisum sativum L. and Vicia faba L. Ann Bot 54:837–841

Gahan P B, Carmignac D F 1987 Determination of vascular tissues in roots of dicotyledonous plants. In: Kolek J, Ciamporova M, Loughman B C (eds) Proc 3rd Int Symp Root Struct Funct. Nitra Czechoslovakia (in press)

Gahan P B, Hurst P R 1976 Effects of ageing on the cell cycle in Zea mays. Ann Bot 40:887–890

Gahan P B, McLean J 1969 Subcellular localization and possible functions of acid β-glycerophosphatases and naphthol esterases in plant cells. Planta 89:126–135

Gahan P B, Rana M A 1985 The quiescent centre and cell determination in roots of Pisum sativum. Ann Bot 56:437–442

Gahan P B, Rana M A, Phillips R 1983 Activation of carboxylesterases in root cortical parenchymal cells of Pisum sativum during xylem induction in vitro. Cell Biochem Funct 1:109–111

Galston A W 1983 Polyamines as modulators of plant development. BioScience 33:328–388

Galun E, Izhar S, Atsmon D 1965 Determination of relativ auxin content in hermaphrodite and andromonoecious *Cucumis sativus* L. Plant Physiol 40:321–326

Gamborg O L, Miller R A, Ojima K 1968 Nutrient requirements of suspension cultures of soybean root cells. Exp Cell Res 50:151–158

Gaspar T, Penel C, Castillo F J, Greppin H 1985 A two-step control of basic and acidic peroxidases and its significance for growth and development. Physiol Plant 64:418–423

Gaspar T, Penel C L, Thorpe T, Greppin H 1982 Peroxidases 1970–1980. A survey of their biochemical and physiological roles in higher plants. Univ Genève, Cent Bot, 324 pp

Gautheret R J 1961 a Action conjugée de l'acide gibbérellin, de la cinétine et de l'acide indoléacétique sur les tissus cultivés in vitro particulièrement sur ceux de Topinambour. C R Acad Sci 253:1381–1385

Gautheret R J 1961 b Action de la lumière et de la température sur la néoformation de racines par des tissus de Topinambour cultivés in vitro. C R Acad Sci 250:2971–2796

Gautheret R J 1966 Factors affecting differentiation in plant tissues grown in vitro. In: Beermann W (ed) Cell differentiation and morphogenesis. Elsevier/North-Holland, Amsterdam, New York, pp 55–95

Gautheret R J 1969 Investigations on the root formation in the tissues of Helianthus tuberosus cultured in vitro. Am J Bot 56:702–717

Gersani M 1987 a The induction of differentiation of organized vessels in a storage organ. Ann Bot 59:31–34

Gersani M 1987 b Vessel differentiation along different tissue polarities. Physiol Plant 70:516–522

Gersani M, Sachs T 1984 Polarity reorientation in beans expressed by vascular differentiation and polar auxin transport. Differentiation 25:205–208

Gilliland M G, Van Staden J, Bruton A G 1984 Studies on the translocation system of guayule (Parthenium argentatum). Protoplasma 122:169–177

Glock H, Gregorius H R 1984 Differentiation – a consequence of idiotype-environment interaction. Biosystems 17:23–34

Goeschl J D, Rappaport L, Pratt H K 1966 Ethylene as a factor regulating the growth of pea epicotyls subjected to physical stress. Plant Physiol 41:877–889

Goldberg R, Catesson A M, Czaninski Y 1983 Some properties of syringaldazine oxidase, a peroxidase specifically involved in lignification processes. Z Pflanzenphysiol 110:267–279

Goldberg R, Lê T, Catesson A M 1985 Localization and properties of cell wall enzyme activities related to the final stages of lignin biosynthesis. J Exp Bot 36:503–510

Goldberg R, Thor V, Dubacq J P, Prat R, Pierron M 1986 Time course of a dedifferentiation process during isolation and culture of protoplasts: comparison between protoplasts from young and mature regions of the mung bean hypocotyl. Plant Sci 46:63–68

Goldsmith M H M, Catealdo D A, Karn J, Brenneman T, Trip P 1974 The rapid nonpolar transport of auxin in the phloem of intact Coleus plants. Planta 116:301–317

Goldsworthy A, Rathore K S 1985 The electrical control of growth in plant tissue cultures: the polar transport of auxin. J Exp Bot 36:1134–1141

Goodwin B C, Cohen M H 1969 A phase-shift model for the spatial and temporal organization of developing systems. J Theor Biol 25:49–107

Gould A R 1978 Diverse pathways of morphogenesis in tissue cultures of the composite Brachycome lineariloba (2n=4). Protoplasma 97:125–135

Gould A R 1984 Control of the cell cycle in cultured plant cells. CRC Crit Rev Plant Sci 1:315–605

Greilhuber J, Loidl J 1983 On regularities of C-banding patterns and their possible cause. In: Brandham P E, Bennett M D (eds) Proc Kew Chromosome Conf. Allen and Unwin London, 344 pp

Gresshoff P M 1978 Phytohormones and growth and differentiation of cells and tissues cultured in vitro. In: Letham D S, Goodwin P B, Higgins T J V (eds) Phytohormones and related compounds – a comprehensive treatise, vol 2. Elsevier/North-Holland, Amsterdam, New York, pp 1–29

Grew N 1682 The anatomy of plants. Johnson Reprint Corp, New York, London (1965), 304 pp

Griffing L R, Ray P M 1979 Dependence of cell wall secretion on calcium. Plant Physiol 63(Suppl):51 (Abstr)

Grisebach H, Hahlbrock K 1974 Enzymology and regulation of flavanoid and lignin biosynthesis in plants and plant cell suspension cultures. In: Runeckles V C, Conn E E (eds) Metabolism and regulation of secondary plant products. Academic Press, New York London, pp 22–52

Guern J, Péaud-Lenoël C eds 1981 Metabolism and molecular activities of cytokinins. Springer, Berlin Heidelberg New York, 352 pp

Guilfoyle T J 1986 Auxin-regulated gene expression in higher plants. CRC Crit Rev Plant Sci 4:247–276

Gunning B E S 1977 Transfer cells and their roles in transport of solutes in plants. Sci Prog 64:539–568

Gunning B E S, Hardham A R 1982 Microtubules. Ann Rev Plant Physiol 33:651–698

Gunning B E S, Pate J S 1969 "Transfer cells". Plant cells with wall ingrowths, specialized in relation to short distance transport of solutes – their occurrence, structure and development. Protoplasma 68:107–133

Habricot Y, Sossountzov L 1984 Distribution, fine structure and possible role of transfer cells in relation to apical dominance in the aquatic fern, Marsilea drummondii. Cytobios 41:191–206

Haddon L, Northcote D H 1975 Quantitative measurement of the course of bean callus differentiation. J Cell Sci 17:11–26

Haddon L, Northcote D H 1976a Correlation of the induction of various enzymes concerned with phenylpropanoid and lignin synthesis during differentiation of bean callus (Phaseolus vulgaris L.). Planta 128:255–262

Haddon L, Northcote D H 1976b The effect of growth conditions and origin of tissue on the ploidy and morphogenetic potential of tissue cultures of bean (Phaseolus vulgaris L.). J Exp Bot 27:189–209

Haddon L, Northcote D H 1976c The influence of gibberellic acid and abscisic acid on cell and tissue differentiation of bean callus. J Cell Sci 20:47–55

Hahlbrock K, Grisebach H 1979 Enzymic controls in the biosynthesis of lignin and flavonoids. Ann Rev Plant Physiol 30:105–130

Hahne G, Hoffmann F 1984 Dimethyl sulfoxide can initiate cell divisions of arrested callus protoplasts by promoting cortical microtubule assembly. Proc Natl Acad Sci USA 81:5449–5453

Hahne G, Hoffmann F 1985 Cortical microtubular lattices: absent from mature mesophyll and necessary for cell division? Planta 166:309–313

Haigler C H, Brown R M, Jr 1986 Transport of rosettes from the Golgi apparatus to the plasma membrane in isolated mesophyll cells of Zinnia elegans during differentiation to tracheary elements in suspension culture. Protoplasma 134:111–120

Hammersley D R H, McCully M E 1980 Differentiation of wound xylem in pea roots in the presence of colchicine. Plant Sci Lett 19:151–156

Hanson E J, Bren P J 1985 Xylem differentiation and boron accumulation in Italian prune flower buds. J Am Soc Hortic Sci 110:566–570

Hardham A R, Gunning B E S 1979 Interpolation of microtubules into cortical arrays during cell elongation and differentiation in roots of Azolla pinnata. J Cell Sci 37:411–442

Hardham A R, McCully M E 1982a Reprogramming of cells following wounding in pea (Pisum sativum L.) roots. I. Cell division and differentiation of new vascular elements. Protoplasma 112:143–151

Hardham A R, McCully M E 1982b Reprogramming of cells following wounding in pea (Pisum sativum L.) roots. II. The effects of caffeine and colchicine on the development of new vascular elements. Protoplasma 112:152–166

Harris N, Chaffey N J 1985 Plasma tubules in transfer cells of pea (Pisum sativum). Planta 165:191–196

Harrison M A, Klein R M 1979 Role of growth regulators in initiation of secondary xylem and phloem cells. Bot Gaz 140:20–24

Hasegawa K, Hashimoto T 1975 Variation in abscisic acid and batatasin content of yam bulbils – effects of stratification and light exposure. J Exp Bot 26:757–764
Hasegawa K, Koreeda M, Mase T 1983 A new growth inhibitor, pisumin, involved in light inhibition of epicotyl growth of dwarf peas. Plant Physiol 72:391–393
Hawkins S W, Phillips R 1983 3-Aminobenzamide inhibits tracheary element differentiation but not cell division in cultured explants of Helianthus tuberosus. Plant Sci Lett 32:221–224
Hayflick L 1965 The limited in vitro lifetime of human diploid cell strains. Exp Cell Res 37:614–636
Hayflick L 1977 The cellular basis for the biology of aging. In: Finch C E, Hayflick L (eds) The biology of aging. Van Nostrand Reinhold, New York, pp 159–188
Heath R L 1980 Initial events in injury to plants by air pollutants. Ann Rev Plant Physiol 31:395–431
Henry Y, Steer M W 1980 A re-examination of the induction of phloem transfer cell development in pea leaves (Pisum sativum). Plant Cell Environ 3:377–380
Hepler P K 1982 Morphogenesis of tracheary elements and guard cells. In: Kiermayer O (ed) Cytomorphogenesis in plants. Springer, Berlin Heidelberg New York, pp 327–347
Hepler P K, Palevitz B A 1974 Microtubules and microfilaments. Ann Rev Plant Physiol 25:309–362
Hepler P K, Rice R M, Terranova W A 1972 Cytochemical localization of peroxidase activity in wound vessel members of Coleus. Can J Bot 50:977–983
Hepler P K, Wayne R O 1985 Calcium and plant development. Ann Rev Plant Physiol 36:397–439
Hepler P K, Wolniak S M 1984 Membranes in the mitotic apparatus: their structure and function. Int Rev Cytol 90:169–238
Hertel R 1983 The mechanism of auxin transport as a model for auxin action. Z Pflanzenphysiol 112:53–67
Hertel R, Flory R 1968 Auxin movement in corn coleoptiles. Planta 82:123–144
Herth W 1985 Plasma-membrane rosettes involved in localized wall thickening during xylem vessel formation of Lepidium sativum L. Planta 164:12–21
Heslop-Harrison J S, Bennett M D 1984 Chromosome order – possible implications for development. In: Sawn A, McGregor H, Ransom R (eds) Programmes for development. J Embryol Exp Morphol 83(Suppl):51–73
Hess T, Sachs T 1972 The influence of a mature leaf on xylem differentiation. New Phytol 71:903–914
Higinbotham N, Davis R F, Mertz S M, Shumway L K 1972 Some evidence that radial transport in maize roots is into living vessels. In: Anderson W P (ed) Ion transport in plants. Academic Press, New York London, pp 493–506
Higuchi T 1985 Biosynthesis of lignin. In: Higuchi T (ed) Biosynthesis and biodegradation of wood components. Academic Press, New York, London, pp 141–160
Hillis W E 1972 Properties of eucalypt woods of importance to the pulp and paper industry. Appita 26:113–122
Hillis W E 1975 Ethylene and extraneous material formation in woody tissues. Phytochemistry 14:2559–2562
Hillis W E 1987 Heartwood and tree exudates. Springer, Berlin Heidelberg New York Tokyo, 268 pp
Himes R H, Burton P R, Gaito J M 1977 Dimethyl sulfoxide-induced self-assembly of tubulin lacking associated proteins. J Biol Chem 252:6222–6228
Hirsch A M, Pengelly W L, Torrey J G 1982 Endogenous IAA levels in boron-deficient and control root tips of sunflower. Bot Gaz 143:15–19
Hirsch A M, Torrey J G 1980 Ultrastructural changes in sunflower root cells in relation to boron deficiency and added auxin. Can J Bot 58:856–866
Hoffman A, Miller A R, Pengelly W L 1985 Characterization of polyphenols in cell walls of cultured Populus trichocarpa tissues. Phytochemistry 24:2685–2687
Hoffmann F, Hahne B, Hahne G 1984 Are cortical microtubules essential for cell formation and cell division by isolated callus protoplasts? In: Dugger W M, Bartnicki-Garcia S (eds) Structure, function, and biosynthesis of plant cell walls. Am Soc Plant Physiol, Rockville, pp 444–448

Holder N 1979 Positional information and pattern formation in plant morphogenesis and a mechanism for the involvement of plant hormones. J Theor Biol 77:195–212

Holta E, Hannonen P, Pispa J, Janne J 1973 Effect of methylglyoxal bis (guanylhydrazone) on polyamine metabolism in normal and regenerating rat liver and rat thymus. Eur J Biochem 136:669–676

Holtzer H 1970 Proliferation and quantal mitosis in differentiation. In: Padykula H A (ed) Control mechanisms in the expression of cellular phenotypes. Academic Press, New York London, pp 69–88

Hooke R 1665 Micrographia. Martyn and Allestry, London, 246 pp

Hoppe D C, McCully M E, Wenzel C L 1986 The nodal roots of Zea: their development in relation to structural features of the stem. Can J Bot 64:2524–2537

Horgan R 1984 Cytokinins. In: Wilkins M B (ed) Advanced plant physiology. Pitman, London, pp 53–75

Houck D F, LaMotte C E 1977 Primary phloem regeneration without concomitant xylem regeneration: its hormone control in Coleus. Am J Bot 64:799–809

Hurst P R, Gahan P B 1975 Turnover of DNA on ageing tissues of Lycopersicon esculentum. Ann Bot 39:71–76

Hurst P R, Gahan P B, Snellen J W 1973 Turnover of labelled DNA in differentiated collenchyma. Differentiation 1:261–266

Hyodo H, Kuroda H, Yang S F 1978 Induction of phenylalanine ammonia-lyase and increase in phenolics in lettuce leaves in relation to the development of russet spotting caused by ethylene. Plant Physiol 62:31–35

Ingold E, Sugiyama M, Komamine A 1987 Secondary cell wall formation: changes in cell wall constituents during the differentiation of single mesophyll cells of Zinnia elegans to tracheary elements. Planta (in press)

Innocenti A M, Avanzi S 1971 Some cytological aspects of the differentiation of metaxylem in the root of Allium cepa. Caryologia 24:283–292

Irving R M 1969 Characterization and role of an endogenous inhibitor in induction of cold hardiness in Acer negundo. Plant Physiol 44:801–805

Iwasaki T, Fukuda H, Shibaoka H 1986 Inhibition of cell division and DNA synthesis by gibberellin in isolated Zinnia mesophyll cells. Plant Cell Physiol 27:717–724

Jackson M B 1985 Ethylene and responses of plants to soil waterlogging and submergence. Ann Rev Plant Physiol 36:145–174

Jacobs M, Gilbert S F 1983 Basal localization of the presumptive auxin transport carrier in pea stem cells. Science 220–1297–1300

Jacobs M, Short T W 1986 Further characterization of the presumptive auxin transport carrier using monoclonal antibodies. In: Bopp M (ed) Plant growth substances. Springer, Berlin Heidelberg New York Tokyo, pp 218–226

Jacobs W P 1952 The role of auxin in differentiation of xylem around a wound. Am J Bot 39:301–309

Jacobs W P 1959 What substance normally controls a given biological process? I. Formulation of some rules. Dev Biol 1:527–533

Jacobs W P 1970 Regeneration and differentiation of sieve-tube elements. Int Rev Cytol 28:239–273

Jacobs W P 1984 Functions of hormones at tissue level of organization. In: Scott T K (ed) Hormonal regulation of development. II. The functions of hormones from the level of the cell to the whole plant. Encycl Plant Physiol, New Ser, vol 10. Springer, Berlin Heidelberg New York, pp 149–171

Jacobs W P, Aloni R 1978 Evidence that IAA does not move basipetally against a concentration gradient in Coleus blumei internodes. Ann Bot 42:989–991

Jacqmard A 1968 Early effects of gibberellic acid on mitotic activity and DNA synthesis in the apical bud of Rudsbeckia bicolor. Physiol Veg 6:409–416

Jacquist C 1972 Réflexions sur les rapports entre tissus secondaire et tissus primaires. Ann Sci Univ Besançon 3 Ser Bot 12:87–89

Jacquoit C 1947 Effet inhibiteur des tannins sur le développement des cultures in vitro du cambium de certains arbres forestiers. C R Acad Sci 225:424–426

Jaffe M J 1973 Thigmomorphogenesis: The response of plant growth and development to mechanical stimulation. With special reference of Bryonia dioica. Planta 114:143–157

Jaffe M J 1980 Morphogenetic responses of plants to mechanical stimuli or stress. BioScience 30:239–243

Jaffe M J, Biro R 1979 Thigmomorphogenesis: The effect of mechanical perturbation on the growth of plants, with special reference to anatomical changes, the role of ethylene, and interaction with other environmental stresses. In: Mussell H, Staples R C (eds) Stress physiology in crop plants. John Wiley, New York, pp 25–69

Jaffe M J, Huberman M, Johnson J, Telewski F W 1985 Thigmomorphogenesis: The induction of callose formation and ethylene evolution by mechanical perturbation in bean stems. Physiol Plant 64:271–279

Jenkins P A 1974 Influence of applied indoleacetic acid and abscisic acid on xylem cell dimensions in Pinus radiata D. Don. In: Bieleski R L, Ferguson A R, Cresswell M M (eds) Mechanisms of regulation of plant growth. R Soc NZ, Wellington, pp 737–742

Jensen W A 1958 The nucleic acid and protein content of root tip cells of Vicia faba and Allium cepa. Exp Cell Res 14:573–583

Johansen D A 1950 Plant embryology. Chronica Botanica, Waltham 305 pp

Johnson A P, Williams G T 1982 Role of DNA breaks and ADP-ribosyl transferase activity in eukaryotic differentiation demonstrated in human lymphocytes. Nature 300:368–370

Johnson D C, Roberts L W 1978 Water stress influences xylogenesis in cultured explants of Lactuca. Phytomorphology 28:207–209

Johri B M 1984 Embryology of angiosperms. Springer, Berlin, 830 pp

Jones D H, Northcote D H 1981 Induction by hormones of phenylalanine ammonia-lyase in bean-cell suspension cultures. Inhibition and superinduction by actinomycin D. Eur J Biochem 116:117–125

Jones M G K, Northcote D H 1972 Multinucleate transfer cells induced in Coleus roots by the rootknot nematode, Meliodogyne arenana. Protoplasma 75:381–395

Jones M G K, Novacky A, Dropkin V H 1974 "Action potentials" in nematode-induced plant transfer cells. Protoplasma 80:401–405

Jones M G K, Payne H L 1978 Cytokinesis in Impatiens balsamina and the effect of caffeine. Cytobios 20:79–91

Jones R L, MacMillan J 1984 Gibberellins. In: Wilkins M B (ed) Advanced plant physiology. Pitman, London, pp 21–52

Kao C H, Yang S F 1982 Light inhibition of the conversion of 1-aminocyclopropane-1-carboxylic acid to ethylene in leaves is mediated through carbon dioxide. Planta 155:261–266

Kato Y 1980 Studies on juice vesicles isolated from mature and immature citrus fruit. I. Callus induction and cell differentiation in Citrus hassaku hort. ex Tanka. J Jap Soc Hortic Sci 49:36–40

Kauss H 1981 Lectins and their physiological role in slime moulds and in higher plants. In: Tanner W, Loewus F A (eds) Plant carbohydrates. Encycl Plant Physiol, New Ser, vol 2. Springer, Berlin Heidelberg New York, pp 627–657

Kauss H, Köhle H, Jeblick W 1983 Proteolytic activation and stimulation by Ca^{2+} of glucan synthase from soybean cells. FEBS Lett 158:84–88

Kay L E, Basile D V 1987 Specific peroxidase isoenzymes are correlated with organogenesis. Plant Physiol 84:99–105

Kessell R H J, Carr A H 1972 The effect of dissolved oxygen concentration on growth and differentiation of carrot (Daucus carota) tissue. J Exp Bot 23:996–1007

Key J L 1969 Hormones and nucleic acid metabolism. Ann Rev Plant Physiol 20:449–474

Khan A, Chauhan Y S, Roberts L W 1986 In vitro studies on xylogenesis in Citrus fruit vesicles. II. Effect of pH of the nutrient medium on the induction of cytodifferentiation. Plant Sci 46:213–216

King G A, Heyes J K 1986 Morphology and cytology of pea embryo during histogenesis. Ann Bot 58:633–640

Kirchner M W 1978 Microtubule assembly and nucleation. Int Rev Cytol 54:1–71

Kirschner H, Sachs T 1978 Cytoplasmic reorientation: An early stage of vascular differentiation. Isr J Bot 27:131–137

Kleiber H, Mohr H 1967 Vom Einfluß des Phytochroms auf die Xylem-differenzierung im Hypokotyl des Senfkeimlings (Sinapis alba L.). Planta 76:85–92

Kobayashi H, Fukuda H, Shibaoka H 1987 Reorganization of actin filaments associated with the differentiation of tracheary elements in Zinnia mesophyll cells. Protoplasma 138:69–71

Köhle H, Heblick W, Poten F, Blaschek W, Kauss H 1985 Chitosan-elicited callose synthesis in soybean cells as a Ca^{2+}-dependent process. Plant Physiol 77:544–551

Kohlenbach H W, Körber M, Lang H, Li L C, Schöpke C 1982a Differentiation in suspension cultures of isolated protoplasts. In: Fujiwara A (ed) Plant tissue culture 1982. Jap Assoc Plant Tissue Cult, Tokyo, pp 95–96

Kohlenbach H W, Körber M, Li L C 1982b Cytodifferentiation of protoplasts isolated from a stem embryo system of Brassica napus to tracheary elements. Z Pflanzenphysiol 107:367–371

Kohlenbach H W, Schmidt B 1975 Cytodifferentiation in the mode of a direct transformation of isolated mesophyll cells to tracheids. Z Pflanzenphysiol 75:369–374

Kohlenbach H W, Schöpke C 1981 Cytodifferentiation to tracheary elements from isolated mesophyll protoplasts of Zinnia elegans. Naturwissenschaften 68:576

Komamine A, Fukuda H 1982 Biochemical mechanism of cytodifferentiation. In: Fujiwara A (ed) Plant tissue culture 1982. Jap Assoc Plant Tissue Cult, Tokyo, pp 91–92

Kordan H A 1964 Vascular elements in juice vesicles of the lemon fruit. Bull Torrey Bot Club 91:271–274

Kuboi T, Yamada Y 1978a Regulation of the enzyme activities related to lignin synthesis in cell aggregates of tobacco cell culture. Biochim Biophys Acta 542:181–190

Kuboi T, Yamada Y 1978b Changing cell aggregation and lignification in tobacco suspension cultures. Plant Cell Physiol 19:437–443

Kubowicz B D, Vanderhoef L N, Hanson J B 1982 ATP-dependent calcium transport in plasmalemma preparations from soybean hypocotyls. Effect of hormone treatments. Plant Physiol 69:187–191

Kucera L J 1977 Modified tracheids adjacent to wound tissue in Pseudowintera colorata (Winteraceae). IAWA Bull N S 1:10–11

Kulaeva O N 1981 Cytokinin action on transcription and translation in plants. In: Guern J, Péaud-Leonoël C (eds) Metabolism and molecular activities of cytokinins. Springer, Berlin Heidelberg New York, pp 218–227

Kuligowski J, Tourte Y 1979 La détermination des cellules apicales chez une ptéridophyte; rôle particulier du génome paternelle. Bull Soc Bot Fr Lett Bot 126:491–505

Kulshresta V K, Chauhan Y S, Roberts L W 1982 In vitro studies on xylogenesis in Citrus fruit vesicles. I. Effects of auxin, cytokinin and citric acid on induction of xylem differentiation. Phytomorphology 32:14–17

Kuo J, Pate J S 1985 Unusual network of internal phloem in the pod mesocarp of cowpea [Vigna unguiculata (L.) Walp. (Fabaceae)]. Ann Bot 55:635–647

Kuroda K, Shimaji K 1983 Traumatic resin canal formation as a marker for xylem growth. For Sci 29:653–659

Kuroda K, Shimaji K 1984a Wound effects on xylem cell differentiation in a conifer. IAWA Bull N S 5:295–305

Kuroda K, Shimaji K 1984b The pinning method for marking xylem growth in hardwood species. For Sci 30:548–554

Kuroda K, Shimaji K 1985 Wound effects on cytodifferentiation in hardwood xylem. IAWA Bull N S 6:107–118

Lauchaud S 1983 Xylogénese chez les Dicotylédones arborescentes. IV. Influence des bourgeons, de l'acide α-indolyl acetique et de l'acide gibberellique sur la réactivation cambiale et al xylogénèse dans les jeunes tiges de Hêtre. Can J Bot 61:1768–1774

Lauchaud S, Bonnemain J L 1984 Seasonal variations in the polar transport pathways and retention sites of [^{3}H]-indole-3-acetic acid in young branches of Fagus sylvatica L. Planta 161:207–215

La Cour L F, Pelc S R 1958 Effect of colchicine on the utilization of labelled thymidine during chromosomal reproduction. Nature 182:506–508

Lai V, Srivastava L M 1976 Nuclear changes during differentiation of xylem vessel elements. Cytobiology 12:220–243

Lamb C J 1983 Molecular approaches to the study of cell differentiation and development in higher plants: the biochemistry of xylem and phloem production. In: Buckingham M E (ed) Biochemistry of cellular regulation. CRC, Boca Raton, pp 145–178

Lambert R G, Linck A J 1958 Effects of high temperature on yield of peas. Plant Physiol 33:347–350

LaMotte C E, Jacobs W P 1962 Quantitative estimation of phloem regeneration in Coleus internodes. Stain Technol 37:63–73

LaMotte C E, Jacobs W P 1963 A role of auxin in phloem regeneration in Coleus internodes. Dev Biol 8:80–98

Lang H, Kohlenbach H W 1982 Differentiation of alkaloid cells in cultures of Macleaya mesophyll protoplasts. Planta Med J Med Plant Res 46:65–128

Larson P R 1976 Procambium versus cambium and protoxylem versus metaxylem in Populus deltoides seedlings. Am J Bot 63:1332–1348

Larson P R 1982 The concept of cambium. In: Baas P (ed) New perspectives in wood anatomy. Nijhoff/Junk, The Hague, pp 85–121

Larson P R 1983 Primary vascularisation and the siting of primordia. In: Dale J E, Milthorpe F L (eds) The growth and functioning of leaves. Cambridge University Press, Cambridge, pp 25–51

Laughlin R G, Munyon R L, Ries S K, Wert V F 1983 Growth enhancement of plants by femtomole doses of colloidally dispersed triacontanol. Science 219:1219–1221

Lavee S, Galston A W 1968 Structural, physiological, and biochemical gradients in tobacco pith tissue. Plant Physiol 43:1760–1768

Leblond C P, Messier B 1958 Renewal of chief cells and goblet cells in the small intestine as shown by autoradiography after injection of thymidine-H^3 into mice. Anat Rec 132:247–260

Lee S, Aronoff S 1967 Boron in plants: a biochemical role. Science 158:798–799

Lee T S, Purse J G, Pryce R J, Horgan R, Wareing P F 1981 Dihydroconiferyl alcohol – a cell division factor from Acer species. Planta 152:571–577

Leopold A C 1955 Auxins and plant growth. University of California Press, Berkeley, Los Angeles, 354 pp

Leopold A C 1972 Ethylene as a plant hormone. In: Kaldewey H, Vardar Y (eds) Hormonal regulation in plant growth and development. Verlag Chemie, Weinheim, pp 245–262

Leopold A C 1977 Modification of growth regulatory action with inorganic solutes. In: Sutte C A (ed) Plant growth regulators. Chemical activity, plant responses, and economic potential. Am Chem Soc, Washington, pp 33–41

Leopold A C, Kriedemann P E 1975 Plant growth and development. 2nd edn. McGraw-Hill, New York, 545 pp

Leopold A C, Noodén L D 1984 Hormonal regulatory system in plants. In: Scott T K (ed) Hormonal regulation of development II. The functions of hormones from the level of the cell to the whole plant. Encycl Plant Physiol, New Ser, vol 10. Springer Berlin Heidelberg, New York, pp 4–22

Lersten N R 1986 Modified clearing method to show sieve tubes in minor veins of leaves. Stain Technol 61:231–234

Letvenuk L J, Peterson R L 1976 Occurrence of transfer cells in vascular parenchyma of Hieracium florentinum roots. Can J Bot 54:1458–1471

Lewis D H 1980 Boron, lignification and the origin of vascular plants – a unified hypothesis. New Phytol 84:209–229

Lieberman M 1979 Biosynthesis and action of ethylene. Ann Rev Plant Physiol 30:533–591

Linstedt D, Reinert J 1975 Occurrence and properties of a cytokinin in tissue cultures of Daucus carota. Naturwissenschaften 62:238–239

Lintilhac P M 1984 Positional controls in meristem development: a caveat and an alternative. In: Barlow P W, Carr D J (eds) Positional controls in plant development. Cambridge University Press, Cambridge, pp 83–106

Lintilhac P M, Vesecky T B 1984 Stress-induced alignment of division plane in plant tissues grown in vitro. Nature 307:363–364

Lipetz J 1962 Calcium and the control of lignification in tissue cultures. Am J Bot 49:460–464

Lipetz J, Garro A J 1965 Ionic effects on lignification and peroxidases in tissue cultures. J Cell Biol 25:109–116
Liphschitz N, Lev-Yadun S, Waisel Y 1981 The annual rhythm of activity of the lateral meristems (cambium and phellogen) in Cupressus sempervirens L. Ann Bot 47:485–496
List A 1963 some observations on DNA content and cell and nuclear volume growth in the developing xylem cells of certain higher plants. Am J Bot 50:320–329
Little C H A, Wareing P F 1981 Control of cambial activity and dormancy in Picea sitchensis by indole-3-acetic acid and abscisic acid. Can J Bot 59:1480–1493
Liu B L L 1984 Abscisic acid induces land form characteristics in Marsilea quadrifolia L. Am J Bot 71:638–655
Liu P B W, Loy J B 1976 Action of gibberellic acid on cell proliferation in the subapical shoot meristem of watermelon seedlings. Am J Bot 63:700–704
Lloyd C W 1986 Microtubules and the cellular morphogenesis of plants. In: Browder L W (ed) Developmental biology. A comprehensive synthesis, vol 2. The cellular basis of morphogenesis. Plenum, New York, pp 31–57
Lovatt C J 1985 Evolution of xylem resulted in a requirement for boron in the apical meristems of vascular plants. New Phytol 99:509–522
Loy J B 1977 Hormonal regulation of cell division in the primary elongation meristems of shoots. In: Rost T L, Gifford E M (eds) Mechanisms and control of cell division. Dowden Hutchinson Ross, Stroudsville, pp 92–110
Lux A 1986 Changes of plastids during xylem differentiation in barley roots. Ann Bot 58:547–550
Luxová M 1986 The hydraulic safety zone at the base of barley roots. Planta 169:465–470
Lynn D G, Chen R H, Manning K S, Wood H N 1987 The structural characterization of endogenous factors from Vinca rosea crown gall tumors that promote cell division of tobacco cells. Proc Natl Acad Sci USA 84:615–619
MacMillan J ed 1980 Hormonal regulation of development I. Molecular aspects of plant hormones. Encycl Plant Physiol, New Ser, vol 9. Springer, Berlin Heidelberg New York, 681 pp
Malawer C L, Phillips R 1979 The cell cycle in relation to induced xylem differentiation. Tritiated thymidine incorporation in cultured tuber explants of Helianthus tuberosus L. Plant Sci Lett 15:47–55
Malpighi M 1675 Anatome plantarum. Martyn and Allestry, London (no pagination)
Marcum J M, Dedman J R, Brinkley B R, Means A R 1978 Control of microtubule assembly-disassembly by calcium-dependent regulatory protein. Proc Natl Acad Sci USA 75:3771–3775
Markland W, Haddon L 1982 An improved technique for the culture of Jersualem artichoke (Helianthus tuberosus L.) explants for use in the study of xylem differentiation. Plant Cell Rep 1:229–231
Marmé D 1983 Calcium transport and function. In: Läuchi A, Bieleski R L (eds) Inorganic plant nutrition. Encycl Plant Physiol, New Ser, vol 15B. Springer, Berlin Heidelberg New York, pp 599–625
Martin R B, Richardson F S 1979 Lanthanides as probes for calcium in biological systems. Q Rev Biol 12:181–209
Martin S M 1980 Environmental factors. B. Temperature, aeration, and pH. In: Staba E J (ed) Plant tissue culture as a source of biochemicals. CRC, Boca Raton, pp 143–148
Masuda H, Fukuda H, Komamine A 1983 Changes in peroxidase isoenzyme patterns during tracheary element differentiation in a culture of single cells isolated from the mesophyll of Zinnia elegans. Z Pflanzenphysiol 112:417–426
Matsuno H, Uritani I 1972 Physiological behavior of peroxidase isozymes in sweet potato root tissue injured by cutting or with black rot. Plant Cell Physiol 13:1091–1101
McCully M E, Canny M J, Van Steveninck R F M 1987 Accumulation of potassium by differentiating metaxylem elements of maize roots. Physiol Plant 69:73–80
McDaniel R G 1982 The physiology of temperature effects on plants. In: Christiansen M N, Lewis C F (eds) Breeding plants for less favorable environments. John Wiley, New York, pp 13–45

McGahan M W 1955 Vascular differentiation in the vegetative shoot of Xanthium chinese. Am J Bot 42:132–140
McGarry A, Gahan P B 1985 A quantitative cytochemical study of UDP-D-glucose dehydrogenase: NAD-oxidoreductase (EC 1.1.1.22) activity during stelar differentiation in Pisum sativum L cv. Meteor. Histochemistry 83:551–554
McKee J M T, Morris G E L 1986 Effects of gibberellic acid and chlormequat on the proportion of phloem and xylem parenchyma in the storage root of carrot (Daucus carota L.). Plant Growth Regul 4:203–211
McLeish J, Sunderland N 1961 Measurements of deoxyribonucleic acid (DNA) in higher plants by Feulgen photometry and chemical methods. Exp Cell Res 34:527–540
McLeod Rd, McLauchlin S M 1974 The development of a quiescent centre in lateral roots. Ann Bot 38:535–544
Meicenheimer R D, Larson P R 1985 Exogenous auxin and *N*-1-naphthyl-phthalamic acid effects on Populus deltoides xylogenesis. J Exp Bot 36:320–329
Meimeth T, Tran Thanh Van K, Marcotte J L, Trinn T H, Clarke A E 1985 Distribution of lectins in tissues, derived callus, and roots of Psophocarpus tetragonolobus (winged bean). Plant Physiol 70:579–584
Meins F, Jr, Binns A N 1979 Cell determination in plant development. BioScience 29:221–225
Milborrow B V 1983 Pathways to and from abscisic acid. In: Addicott F T (ed) Abscisic acid. Praeger, New York, pp 79–111
Milborrow B V 1984 Inhibitors. In: Wilkins M B (ed) Advanced plant physiology. Pitman, London, pp 76–110
Miller A R, Crawford D L, Roberts L W 1985 Lignification and xylogenesis in Lactuca pith explants cultured in vitro in the presence of auxin and cytokinin: a role for endogenous ethylene. J Exp Bot 36:110–118
Miller A R, Pengelly W L, Roberts L W 1984 Induction of xylem differentiation in Lactuca by ethylene. Plant Physiol 75:1165–1166
Miller A R, Roberts L W 1982 Regulation of tracheary element differentiation by exogenous L-methionine in callus of soya bean cultivars. Ann Bot 50:111–116
Miller A R, Roberts L W 1984 Ethylene biosynthesis and xylogenesis in Lactuca pith explants cultured in vitro in the presence of auxin and cytokinin: the effect of ethylene precursors and inhibitors. J Exp Bot 35:691–698
Miller A R, Roberts L W 1986 A relationship between tracheary element formation and lignification in soybean (Glycine max L var Wayne) callus cultures? Can J Bot 64:2716–2718
Miller C O, Skoog F, von Saltza HH, Okumura F S, Strong F H 1956 Isolation, structure, and synthesis of kinetin, a substance promoting cell division. J Am Chem Soc 78:1175
Miller H A, Wetmore R H 1945 Studies in the developmental anatomy of Phlox drummondii Hook I. The embryo. Am J Bot 32:588–599
Minocha S C 1979 The role of auxin and abscisic acid in the induction of cell division in Jerusalem artichoke tuber tissue cultured in vitro. Z Pflanzenphysiol 92:431–441
Minocha S C 1984 The role of benzyladenine in the differentiation of tracheary elements in Jerusalem artichoke tuber explants cultured in vitro. J Exp Bot 35:1003–1015
Minocha S C, Dibona S 1979 Effect of auxin and abscisic acid on RNA and protein synthesis prior to the first cell division in Jerusalem artichoke tuber tissue cultured in vitro. Z Pflanzenphysiol 92:367–374
Minocha S C, Halperin W 1974 Hormones and metabolites which control tracheid differentiation with or without concomitant effects on growth in cultured tuber tissue of Helianthus tuberosus L. Planta 116:319–331
Minocha S C, Halperin W 1976 Enzymatic changes and lignification in relation to tracheid differentiation in cultured tuber tissue of Jerusalem artichoke (Helianthus tuberosus). Can J Bot 54:79–89
Mita T, Katsumi M 1986 Gibberellin control of microtubule arrangement in the mesocotyl epidermal cells of the D5 mutant of Zea mays L. Plant Cell Physiol 27:651–659
Mita T, Shibaoka H 1984a Gibberellin stabilizes microtubules in onion leaf sheath cells. Protoplasma 119:100–109
Mita T, Shibaoka H 1984b Effects of S-3307, an inhibitor of gibberellin biosynthesis, on swelling of leaf sheath cells and on the arrangement of cortical microtubules in onion seedlings. Plant Cell Physiol 25:1531–1539

Mitchison G J 1981 The polar transport of auxin and vein pattern in plants. Philos Trans R Soc London Ser B 295:461–471

Mizuno K, Komamine A 1978 Isolation and identification of substances inducing formation of tracheary elements in cultured carrot-root slices. Planta 138:59–62

Mizuno K, Komamine A, Shimokoriyama M 1971 Vessel element formation in cultured carrot-root phloem slices. Plant Cell Physiol 12:823–830

Montavon M, El Toukhi M, Auderset G, Greppin G 1987 Effect of photoperiod and electrical potentials applied to petioles on the glucose-6-phosphate dehydrogenase activity used as a marker of floral induction in shoot apices of spinach. Ann Bot 60:225–230

Montavon M, Greppin H 1986 Apical development of spinach and application of imposed electrical potential. Saussurea 17:85–91

Morgan P W, Fowler J L 1972 Ethylene: Modifications of peroxidase activity and isozyme complement in cotton (Gossypium hirsutum L.). Plant Cell Physiol 13:727–736

Mori T 1972 Tissue differentiation and development in rice roots. Bull Fac Agric Hirosaki Univ 18:236–304

Morris D A 1980 The influence of small direct electric currents on the transport of auxin in intact plants. Planta 150:431–434

Morris D A, Kadir G O 1972 Pathways of auxin transport in the intact pea seedling (Pisum sativum L.). Planta 107:171–182

Morris D A, Kadir G O, Barry A J 1973 Auxin transport in intact pea seedlings (Pisum sativum L.): the inhibition of transport by 2,3,5-triiodobenzoic acid. Planta 110:173–182

Morris D A, Thomas A G 1978 A microautoradiographic study of auxin transport in the stem of intact pea seedlings (Pisum sativum L.). J Exp Bot 29:147–158

Mozaffari F D S, Gahan P B 1978 Chromosome aberrations and ageing root meristems. Ann Bot 42:1161–1170

Murashige T, Skoog F 1962 A revised medium for rapid growth and bioassays with tobacco tissue cultures. Physiol Plant 15:473–497

Nagahashi G, Thomson W W, Leonard R T 1974 The Casparian strip as a barrier to the movement of lanthanum in corn roots. Science 183:670–671

Nägl W 1973 The mitotic and endomitotic nuclear cycle in Allium carnatium IV H^3-uridine incorporation. Chromosoma 44:203–212

Nägl W 1978 Endopolyploidy and polyteny in differentiation and evolution. Elsevier/North Holland, Amsterdam New York, 283 pp

Naik G G 1965 Studies on the effects of temperature on the growth of plant tissues. MSc thesis, Univ Edinburgh

Naylor A W 1984 Functions of hormones at the organ level of organization. In: Scott T K (ed) Hormonal regulation of development II. The functions of hormones from the level of the cell to the whole plant. Encycl Plant Physiol, New Ser, vol 10. Springer, Berlin Heidelberg New York, pp 172–218

Nelson N D 1978 Xylem ethylene, phenol-oxidizing enzymes, and nitrogen and heartwood formation in walnut and cherry. Can J Bot 56:626–634

Nelson N D, Hillis W E 1978a Genetic and biochemical aspects of kino vein formation in Eucalyptus. II. Hormonal influence on kino formation in E. regnans. Aust For Res 8:83–91

Nelson N D, Hillis W E 1978b Association between altitude and xylem ethylene levels in Eucalyptus pauciflora. Aust For Res 8:69–73

Nelson N D, Hillis W E 1978c Ethylene and tension wood formation in Eucalyptus gomphocephala. Wood Sci Technol 12:309–315

Nelson N D, Rietveld W J, Isebrands J C 1981 Xylem ethylene production in five black walnut families in the early stages of heartwood formation. For Sci 27:537–543

Newcomb W, Peterson R L 1979 The occurrence and ontogeny of transfer cells associated with lateral root nodules in Leguminosae. Can J Bot 57:2583–2602

Newman I V 1956 Patterns in meristems of vascular plants. I. Cell partition in living apices and in cambial zone in relation to concepts of initial cells and apical cells. Phytomorphology 6:1–19

Niranjana R, Shivamurthy G R 1987 Graniferous tracheary elements in the haustorium of Osyris aborea Wall (Santalaceae). Ann Bot 59:237–243

Nitsch C 1968 Effects of growth substances on the induction of flowering of a short-day plant in vitro. In: Wightman F, Setterfield C (eds) Biochemistry and physiology of plant growth substances. Runge Press, Ottawa, pp 1385–1398

Nitsch J P, Nitsch C 1960 Le problème de l'action des auxines sur la division cellulaire: présence d'un cofacteur de division dans le tubercule de Topinambour. Ann Physiol Veg 2:261–268

Nix L E, Wodzicki T J 1974 The radial distribution and metabolism of IAA-^{14}C in Pinus echinata in relation to wood formation. Can J Bot 52:1349–1355

Northcote D H 1982a Macromolecular aspects of cell wall differentiation. In: Boulter D, Parthier B (eds) Nucleic acids and proteins in plants I. Structure, biochemistry and physiology of proteins. Encycl Plant Physiol, New Ser, vol 14A. Springer, Berlin Heidelberg New York, pp 637–655

Northcote D H 1982b Control of enzyme activity during plant cell development. In: Yeoman M M, Truman D E S (eds) Differentiation in vitro. Cambridge University Press, Cambridge, pp 49–64

Northcote D H 1984 Control of cell wall assembly during differentiation. In: Dugger W M, Bartinicki-Garcia S (eds) Structure, function, and biosynthesis of plant cell walls. Am Soc Plant Physiol, Rockville, pp 222–234

Northcote D H 1985a Cell organelles and their function in biosynthesis of cell wall components: control of cell wall assembly during differentiation. In: Higuchi T (ed) Biosynthesis and biodegradation of wood components. Academic Press, New York London, pp 87–108

Northcote D H 1985b Control of cell wall formation during growth. In: Brett C T, Hillman J R (eds) Biochemistry of plant cell walls. Cambridge University Press, Cambridge, pp 177–197

Nowacki J, Bandurski R S 1980 Myo-inositol esters of indole-3-acetic acid as seed auxin precursors of Zea mays L. Plant Physiol 65:422–427

Nurse P, Bissett Y 1981 Gene required in G_1 for commitment to cell cycle and in G_2 for control of mitosis in fission yeast. Nature 292:558–560

Nuti Ronchi V, Caligo M A, Nozzolini M, Luccarini G 1984 Stimulation of carrot somatic embryogenesis by proline and serine. Plant Cell Rep 3:210–214

O'Brien T P 1981 The primary xylem. In: Barnett J R (ed) Xylem cell development. Castle House, Tunbridge Wells, pp 14–16

Olien W C, Bukovac M J 1982 Ethephon-induced gummosis in sour cherry (Prunus cerasus L.). 1. Effect on xylem function and shoot water status. Plant Physiol 70:547–555

Parr A J, Loughman B C 1983 Boron and membrane function in plants. In: Robb D A, Pierpoint W S (eds) Metals and micronutrients: uptake and utilization by plants. Academic Press, New York London, pp 87–107

Parsons L R 1982 Plant responses to water stress. In: Christiansen M N, Lewis C F (eds) Breeding plants for less favorable environments. John Wiley, New York, pp 175–192

Pate J S, Gunning B E S 1969 Vascular transfer cells in angiosperm leaves. A taxonomic and morphological survey. Protoplasma 68:135–156

Pate J S, Gunning B E S 1972 Transfer cells. Ann Rev Plant Physiol 23:173–196

Paton D M, Dhawan A K, Willing R R 1980 Effect of Eucalyptus growth regulators on the water loss from plant leaves. Plant Physiol 66:254–256

Pearce D, Miller A R, Pharis R P 1985 Xylogenesis in lettuce pith cultures: A possible role for gibberellins. Plant Physiol 77(Suppl):79 (Abstr)

Pearce D, Miller A R, Roberts L W, Pharis R P 1987 Gibberellin mediated synergism of xylogenesis in lettuce pith cultures. Plant Physiol 84:1121–1125

Pelc S R 1972 Metabolic DNA in ciliated protozoa, salivary gland chromosomes and mammalian cells. Int Rev Cytol 32:327–355

Pelc S R, Gahan P B 1959 The incorporation of labelled thymidine in the seminal vesicle of the mouse. Nature 183:335–338

Pelc S R, La Cour L F 1959 The incorporation of ^{3}H-thymidine in newly differentiated nuclei of roots of Vicia faba. Experimentia 15:131–133

Penel C, Darimont E, Greppin H, Gaspar Th 1979 Rôle du calcium dans l'association de peroxydases à des membranes de racines de Lentille. C R Acad Sci 289:529–532

Penel C, Sticher L, Kevers C, Gaspar Th, Greppin H 1984 Calcium controlled peroxidase secretion by sugar-beet cells: effect of ionophores in relation to organogenesis. Biochem Physiol Pflanz 179:173–180

Peterson C A, Fletcher R A 1973 Lactic acid clearing and fluorescent staining for demonstration of sieve tubes. Stain Technol 48:23–27

Peterson L R, Yeung E C 1975 Ontogeny of phloem transfer cells in Hieracium floribundum. Can J Bot 53:2745–2758

Phelps J E, McGinnes E A, Jr, Saniewski M, Pieniazek J, Smoliński M 1980 Some anatomical observations on the effect of morphactin IT 3456 and ethrel on wood formation in Salix fragilis L. IAWA Bull N S 1–2: 76–82

Phelps J E, Nelson N D, Rietveld W J, McGinnes E A, Jr 1983 Rates of ethylene production by parenchyma cells in black walnut sapwood. Wood Fiber Sci 15:23–27

Phelps J E, Saniewski M, Smoliński M, Pieniazek J, McGinnes E A, Jr 1974 A note on the structure of morphactin induced wood in two coniferous species. Wood Fiber 6:13–17

Philipson W R, Ward J M 1965 The ontogeny of the vascular cambium in the stem of seed plants. Biol Rev Cambridge Philos Soc 40:534–579

Philipson W R, Ward J M, Butterfield B G 1971 The vascular cambium. Chapman and Hall, London, 182 pp

Phillips R 1980 Cytodifferentiation. In: Vasil I K (ed) Recent advances in plant tissue culture. Int Rev Cytol (Suppl) 11A. Academic Press, New York, London, pp 55–70

Phillips R 1981 a Characterization of mitotic cycles preceding xylogenesis in cultured explants of Helianthus tuberosus L. Ann Bot 47:785–792

Phillips R 1981 b Direct differentiation of tracheary elements in cultured explants of gamma-irradiated tubers of Helianthus tuberosus. Planta 153:262–266

Phillips R 1987 Effects of sequential exposure to auxin and cytokinin on xylogenesis in cultured explants of Jerusalem artichoke (Helianthus tuberosus L.). Ann Bot 59:245–250

Phillips R, Arnott S M 1983 Studies on induced tracheary element differentiation in cultured tissues of tubers of the Jerusalem artichoke, Helianthus tuberosus. Histochm J 15:427–436

Phillips R, Arnott S M, Kaplan S E 1981 Antibiotics in plant tissue culture: rifampicin effectively controls bacterial contaminants without affecting the growth of short term explant cultures of Helianthus tuberosus. Plant Sci Lett 21:235–240

Phillips R, Dodds J H 1977 Rapid differentiation of tracheary elements in cultured explants of Jerusalem artichoke. Planta 135:207–212

Phillips R, Hawkins S W 1985 Characteristics of the inhibition of induced tracheary element differentiation by 3-aminobenzamide and related compounds. J Exp Bot 36:119–128

Phillips R, Press M C, Eason A 1987 Polyamines in relation to cell division and xylogenesis in cultured explants of Helianthus tuberosus: lack of evidence for growth-regulatory action. J Exp Bot 38:164–172

Phillips R, Torrey J G 1973 DNA synthesis, cell division and specific cytodifferentiation in cultured pea root cortical explants. Dev Biol 31:336–347

Phillips R, Torrey J G 1974 DNA levels in differentiating tracheary elements. Dev Biol 39:322–325

Pizzolato T D 1982 Anatomical modification by 2,4 DB of vascular cambium and secondary xylem in soybean internodes. Can J Bot 60:2142–2146

Pizzolato T D, Regehr D L 1981 Anatomical effects of 2,4-DB on tomato internodes. Can J Bot 59:1749–1760

Poovaiah B W, Leopold A C 1976a Effects of inorganic salts on tissue permeability. Plant Physiol 58:182–185

Poovaiah B W, Leopold A C 1976b Effects of inorganic solutes on the binding of auxin. Plant Physiol 58:783–785

Porandowski J, Rakowski K, Wodzicki T J 1982 Apical control of xylem formation in the pine stem. II. Response of differentiating tracheids. Acta Soc Bot Pol 51:203–214

Prazak J A, Peterson C A 1987 A rapid fluorescence method for observing phloem regeneration in Coleus blumei. Stain Technol 62:276–278

Preston R D 1981 Introduction. In: Barnett J R (ed) Xylem cell development. Castle House, Tunbridge Wells, pp 1–13

Quirk J T, Smith D M, Freese F 1975 Effect of mechanical stress on growth and anatomical structure of red pine (Pinus resinosa Ait.): torque stress. Can J For Res 5:691–699

Raghavan V 1986 Embryogenesis in angiosperms. A developmental and experimental study. Cambridge University Press, Cambridge, 303 pp

Raghothama K G, Mizrahi Y, Poovaiah B W 1985 Effect of calmodulin antagonists on auxin-induced elongation. Plant Physiol 79:28–33

Rana M A, Gahan P B 1981 Early cytochemical identification of cortical cells stimulated to form xylem elements in vitro in roots of Pisum sativum. Proc R Microsc Soc 16:246

Rana M A, Gahan P B 1982 Initiation of determination of stelar elements in roots of Pisum sativum. Ann Bot 50:757–762

Rana M A, Gahan P B 1983a A quantitative cytochemical study of determination for xylem-element formation in response to wounding in roots of Pisum sativum L. Planta 157:307–316

Rana M A, Gahan P B 1983b Determination of fascicular, interfascicular and cork cambia in dicotyledonous plants. Saussurea 14:51–60

Rao K S, Catesson A M 1987 Changes in the membrane components of nondividing cambial cells. Can J Bot 65:246–254

Rappaport L 1980 Plant growth hormones: internal control points. Bot Gaz 141:125–130

Rathore K S, Goldsworthy A 1985 Electrical control of growth in plants tissues. BioTechnology 3:253–254

Reeve R M 1948 Late embryogeny and histogenesis in Pisum. Am J Bot 35:591–601

Rayle D L, Cleland R 1970 Enhancement of wall loosening and elongation by acid solutions. Plant Physiol 46:250–253

Reinhold L, Sachs T, Vislovska L 1972 The role of auxin in thigmotropism. In: Carr D J (ed) Plant growth substances 1970. Springer, Berlin Heidelberg, New York, pp 731–737

Reiss H D, Herth W 1978 Visualization of the Ca^{2+} gradient in growing pollen tubes of Lilium longiflorum with chlorotetracycline. Protoplasma 97:373–377

Reiss H D, Herth W 1979 Calcium ionophore A 23187 affects localized wall secretion in the tip region of pollen tubes of Lilium longiflorum. Planta 145:225–232

Reynolds T L 1986 Effects of chlorsulfuron, valine and isoleucine, on division and tracheary element differentiation in cell suspension cultures of Solanum carolinense L. J Plant Physiol 125:179–184

Rhodes M J C, Wooltorton L S C 1973 Stimulation of phenolic acid and lignin biosynthesis in Swede root tissue by ethylene. Phytochemistry 12:107–118

Rier J P, Jr 1976 Ozone and vascular tissue differentiation in plants. US Environ Protect Agency, Washington (EPA-600/3-76-068)

Rikin A, Atsmon D, Gitler C 1980 Chilling injury in cotton (Gossypium hirsutum L.): effects of antimicrotubular drugs. Plant Cell Physiol 21:829–837

Robbertse P J, McCully M E 1979 Regeneration of vascular tissue in wounded pea roots. Planta 145:167–173

Roberts L W 1969 The initiation of xylem differentiation. Bot Rev 35:201–250

Roberts L W 1976 Cytodifferentiation in plants. Xylogenesis as a model system. Cambridge University Press, Cambridge, 160 pp

Roberts L W 1983 The influence of physical factors on xylem differentiation. In: Dodds J H (ed) Tissue culture of trees. Croom Helm, London Canberra, pp 88–102

Roberts L W, Baba S 1968 Effect of proline on wound vessel member formation. Plant Cell Physiol 9:353–360

Roberts L W, Baba S 1978 Exogenous methionine as a nutrient supplement for the induction of xylogenesis in lettuce pith explants. Ann Bot 42:375–379

Roberts L W, Baba S 1982 Glycerol and myo-inositol as carbon sources for the induction of xylogenesis in explants of Lactuca. Can J Bot 60:1204–1206

Roberts L W, Baba S 1987 Evidence that auxin-induced xylogenesis in Lactuca explants requires calmodulin. Environ Exp Bot 27:289–295

Roberts L W, Baba S, Shiraishi T, Miller A R 1982 Progress in cytodifferentiation under in vitro conditions. In: Fujiwara A (ed) Plant tissue culture 1982. Jap Assoc Plant Tissue Cult, Tokyo, pp 87–90

Roberts L W, Miller A R 1982 Ethylene and xylem differentiation. What's New Plant Physiol 13:13–16

Roberts L W, Stiff C M, Baba S 1984 Effects of six different agars on tracheary element differentiation in explants of Lactuca. Plant Tissue Cult Lett 1:22–24

Robitaille H A 1975 Stress ethylene production in apple shoots. J Am Soc Hortic Sci 100:524–527

Robitaille H A, Leopold A C 1974 Ethylene and the regulation of apple stem growth under stress. Physiol Plant 32:301–304

Roland J C 1978 Early differences between radial walls and tangential walls of actively growing cambial zone. IAWA Bull N S 1:7–10

Rost T L, Reynolds T L 1985 The reversal of chlorsulfuron-induced inhibition of mitotic entry by isoleucine and valine. Plant Physiol 77:481–482

Rubery P H, Fosket D E 1969 Changes in phenylalanine ammonia-lyase during xylem differentiation in Coleus and soybean. Planta 87:54–62

Rubery P H, Northcote D H 1968 Site of phenylalanine ammonia-lyase activity and synthesis of lignin during xylem differentiation. Nature 219:1230–1234

Rugenstein S R 1982 Tissue response to palladium microprobe, as observed in Gossypium hirsutum L. (Malvaceae). Am J Bot 69:519–528

Sabnis D D, Hart J W 1982 Microtubule proteins and P-proteins. In: Boulter D, Parthier B (eds) Nucleic acids and proteins in plants. I. Structure, biochemistry and physiology of proteins. Encycl Plant Physiol, New Ser, vol 14A. Springer, Berlin Heidelberg New York, pp 401–437

Sachs T 1968 The role of the root in the induction of xylem differentiation in pea. Ann Bot 32:391–399

Sachs T 1969 Polarity and the induction of organized vascular tissues. Ann Bot 33:263–275

Sachs T 1972a A possible basis for apical organization in plants. J Theor Biol 37:353–361

Sachs T 1972b The induction of fibre differentiation in peas. Ann Bot 36:189–197

Sachs T 1975 The induction of transport channels by auxin. Planta 127:201–206

Sachs T 1977 Control of vascular differentiation. In: Sutcliffe J F, Pate J S (eds) The physiology of the garden pea. Academic Press, New York London, pp 213–233

Sachs T 1978 Patterned differentiation in plants. Differentiation 11:65–73

Sachs T 1981 The control of the patterned differentiation of vascular tissues. Adv Bot Res 9:151–262

Sachs T 1984 Axiality and polarity in vascular plants. In: Barlow P W, Carr D J (eds) Positional controls in plant development. Cambridge University Press, Cambridge, pp 193–224

Sachs T 1986 Cellular patterns determined by polar transport. In: Bopp M (ed) Plant growth substances. Springer, Berlin Heidelberg New York, Tokyo, pp 231–235

Sachs T, Cohen D 1982 Circular vessels and the control of vascular differentiation in plants. Differentiation 21:22–26

Saks Y, Aloni R 1985 Polar gradients of tracheid number and diameter during primary and secondary xylem development in young seedlings of Pinus pinca L. Ann Bot 56:771–778

Saks Y, Feigenbaum P, Aloni R 1984 Regulatory effect of cytokinin on secondary xylem fiber formation in an in vivo system. Plant Physiol 76:638–642

Salleo S, Lo Gullo M A, Siracusano L 1984 Distribution of vessel ends in stems of some diffuse- and ring-porous trees: the nodal regions as "safety zones" of the water conducting system. Ann Bot 54:543–552

Samimy C 1978 Influence of cobalt on soya bean hypocotyl growth and its ethylene evolution. Plant Physiol 62:1005–1006

Samimy C, LaMotte C E 1976 Anomalous temperature dependence of seedling development in some soya bean (Glycine max L. Merr) cultivars. Plant Physiol 58:786–789

Sampson M, Davis D D 1966 Metabolically labile DNA in mitotic and non-mitotic cells of Zea mays. Life Sci 5:1239–1247

Saunders M J, Hepler P K 1981 Localization of membrane-associated calcium during cytokinin treatment in Funaria using chlorotetracycline. Planta 152:272–281

Saunders M J, Hepler P K 1982 Calcium ionophore A 23187 stimulates cytokinin-like mitosis in Funaria. Science 217:943–945

Saussay R, Gautheret R 1974 Action de la lumière, de la température et de l'acide gibbérellique sur la production de formations cribovasculaires par les tissus de rhizomes de Topinambour cultiveés in vitro. C R Acad Sci 279:1871–1876

Savidge R A 1983a The role of plant hormones in higher plant cellular differentiation. I. A critique. Histochem J 15:437–445

Savidge R A 1983b The role of plant hormones in higher plant cellular differentiation. II. Experiments with the vascular cambium, and sclereid and tracheid differentiation in the pine, Pinus contorta. Histochem J 15:447–466

Savidge R A 1987 Dihydroconiferyl alcohol in developing xylem of Pinus contorta. Phytochemistry 26:93–94

Savidge R A, Farrar J L 1984 Cellular adjustments in the vascular cambium leading to spiral grain formation in conifers. Can J Bot 62:2872–2879

Savidge R A, Mutumba G M C, Heald J K, Wareing P F 1983 Gas chromatography-mass spectroscopy identification of 1-aminocyclopropane-1-carboxylic acid in compression wood vascular cambium of Pinus contorta Dougl. Plant Physiol 71:434–436

Savidge R A, Wareing P F 1981 a A tracheid-differentiation factor from pine needles. Planta 153:395–404

Savidge R A, Wareing P F 1981 b Plant-growth regulators and the differentiation of vascular elements. In: Barnett J R (ed) Xylem cell cevelopment. Castle House, Tunbridge Wells, pp 192–233

Savidge R A, Wareing P F 1984 Seasonal cambial activity and xylem development in Pinus contorta in relation to endogenous indol-3-yl-acetic acid and (S)-abscisic acid levels. Can J For Res 14:676–682

Schmid R 1976 The elusive cambium. Another terminological contribution. IAWA Bull 4:51–58

Schneider B, Herth W 1986 Distribution of plasma membrane rosettes and kinetics of cellulose formation in xylem development of higher plants. Protoplasma 131:142–152

Schröder J, Betz B, Hahlbrock K 1977 Messenger RNA controlled increase of phenylalanine ammonia-lyase activity in parsley. Plant Physiol 60:440–445

Schulz A 1986 a Wound phloem in transition to bundle phloem in primary roots of Pisum sativum L. I. Development of bundle-leaving wound-sieve tubes. Protoplasma 130:12–26

Schulz A 1986 b Wound phloem in transition to bundle phloem in primary roots of Pisum sativum L. II. The plasmatic contact between wound-sieve tubes and regular phloem. Protoplasma 130:27–40

Schulz A 1987 Sieve element differentiation and fluoresceine translocation in wound-phloem of pea roots after complete severance of the stele. Planta 170:289–299

Seibert M, Kadkade P G 1980 Environmental factors. A. Light. In: Stabe E J (ed) Plant tissue culture as a source of biochemicals. CRC, Boca Raton, pp 123–141

Shah J J, Mair G M, Thanki Y J, Kothari I L 1983 Isolated sieve tubes in the fruit of Coccinia grandis (L.) Voight (Cucurbitaceae). Ann Bot 51:251–253

Shain L, Hillis W E 1973 Ethylene production in xylem of Pinus radiata in relation to heartwood formation. Can J Bot 51:1331–1335

Sham M H, Northcote D H 1987 Transcription and translation of phloem protein (PP2) during phloem differentiation in Cucurbita maxima. Planta 170:392–399

Sheldrake A R 1973 The production of hormones in higher plants. Biol Rev 48:509–529

Sheldrake A R, Northcote D H 1968 a The production of auxin by autolysing tissues. Planta 80:227–236

Sheldrake A R, Northcote D H 1968 b Some constituents of xylem sap and their possible relationship to xylem differentiation. J Exp Bot 19:681–689

Sheriff D W 1983 Control by indole-3-acetic acid of wood production in Pinus radiata D. Don segments in culture. Aust J Plant Physiol 10:131–135

Shibaoka H 1974 Involvement of wall microtubules in gibberellin promotion and kinetin inhibition of stem elongation. Plant Cell Physiol 15:255–263

Shih L M, Kaur-Sawhney R, Fuhrer J, Samanta S, Galston A W 1982 Effects of exogenous 1,3-diaminopropane and spermidine on senescence of oat leaves. Plant Physiol 70:1592–1596

Shininger T L 1971 The regulation of cambial division and secondary xylem differentiation in Xanthium by auxins and gibberellin. Plant Physiol 47:417–422

Shininger T L 1975 a Is DNA synthesis required for the induction of differentiation in quiescent root cortical parenchyma? Dev Biol 45:137–150

Shininger T L 1975 b The morphological, anatomical and cytological effects of gibberellins. In: Krishnamoorthy H N (ed) Gibberellins and plant growth. Wiley Eastern, New Dehli, pp 203–224

Shininger T L 1979 a Xylem and nonxylem cell formation in cytokinin-stimulated root tissue. Quantitative analysis of temperature effects. Proc Natl Acad Sci USA 76:1921–1923

Shininger T L 1979 b The control of vascular development. Ann Rev Plant Physiol 30:313–337

Shininger T L, Torrey J G 1974 The roles of cytokinins in the induction of cell division and cytodifferentiation in pea root cortical tissue in vitro. In: Bieleski R L, Ferguson A R, Cresswell M M (eds) Mechanism of regulation of plant growth. R Soc N Z, Wellington, pp 721–728

Siebers A M 1971 a Initiation of radial polarity in the interfascicular cambium of Ricinus communis L. Acta Bot Neerl 20:211–220

Siebers A M 1971 b Differentiation of isolated interfascicular tissue of Ricinus communis. Acta Bot Neerl 20:344–355

Siebers A M 1972 Vascular bundle differentiation and cambial development in cultured tissue blocks excised from the embryo of Ricinis communis L. Acta Bot Neerl 21:327–342

Simpson S F, Torrey J G 1977 Hormonal control of deoxyribonucleic acid and protein syntheses in pea root cortical explants. Plant Physiol 59:4–9

Sircar S M, Chakraverty R 1960 The effect of gibberellic acid on jute (Corchorus capsularis). Sci Cult 26:141–143

Sjolund R D 1987 Sieve elements in tissue culture: freeze-fracture and isolation. Abstr Proc XIV Int Bot Congr, Berlin, p 227

Smart C C, Amrhein N 1985 The influence of lignification on the development of vascular tissue in Vigna radiata L. Protoplasma 124:87–95

Smith D L 1981 Cotyledons of the Leguminosae. In: Polhill R M, Raven P H (eds) Advances in legume systematics. R Bot Gard, Kew, Surrey, pp 927–940

Smith F, Montgomery R 1959 The chemistry of plant gums and mucilages. Reinhold, New York, p 106, 166, pp 264–269

Snijman D A 1972 Tracheary element differentiation in in vitro grown Nicotiana tabacum L. callus. MS thesis, Univ Natal, Pietermaritzburg

Sorokin H P, Mathus S N, Thimann K V 1962 The effects of auxins and kinetin on xylem differentiation in the pea epicotyl. Am J Bot 49:444–454

Sorokin H P, Sommer A L 1929 Changes in the cells and tissues of root tips induced by the absence of calcium. Am J Bot 16:23–39

Sorokin H P, Sommer A L 1940 Effects of calcium deficiency upon the roots of Pisum sativum. Am J Bot 27:308–318

Sorokin H P, Thimann K V 1964 The histological basis for inhibition of axillary buds in Pisum sativum and the effects of auxins and kinetin on xylem development. Protoplasma 59:326–350

Souèges R 1920 Embryogénie des Solancées. Dévelopment de l'embryon chez les Nicotiana. C R Acad Sci 170:1125–1127

Stant M Y 1961 The effect of gibberellic acid on fibre-cell length. Ann Bot 25:453–462

Stant M Y 1963 The effect of gibberellic acid on cell width and the cell-wall of some phloem fibers. Ann Bot 27:185–190

Starbuck C J, Phelps J E 1986 Induction of compression wood in rooted cuttings of Pseudotsuga menziesii (Mirb.) Franco by indole-3-acetic acid. IAWA Bull N S 7:13–16

Starbuck C J, Roberts A N 1983 Compression wood in rooted cuttings of Douglas-fir. Physiol Plant 57:371–374

St Aubin G, Canny M J, McCully M E 1986 Living vessel elements in the late metaxylem of sheathed maize roots. Ann Bot 58:577–588

Sterling C 1946 Growth and vascular development in the shoot apex of Sequoia sempervirens (Lamb.) Endl. III Cytological aspects of vascularization. Am J Bot 33:35–45

Steward F C, Mapes M O, Mears K 1958 Growth and organized development of cultured cells. II Organization in cultures grown from freely suspended cells. Am J Bot 45:705–708

Stoddard F L, McCully M E 1980 Effects of excision of stock and scion organs on the formation of the graft union in Coleus: histological study. Bot Gaz 141:401–412

Stoddart J L, Venis M A 1980 Molecular and subcellular aspects of hormone action. In: MacMillan J (ed) Hormonal regulation of development I Molecular aspects of plant hormones. Encycl Plant Physiol, New Ser, vol 9. Springer, Berlin Heidelberg New York, pp 445–510

Street H E, Öpik H 1984 The physiology of flowering plants: their growth and development, 3rd edn. Edward Arnold, London, 279 pp

Sudberry P E, Goodeu A R, Carter B L A 1980 Genes which control cell proliferation in the yeast Saccharomyces cerevisiae. Nature 288:401–404

Sugiyama M, Fukuda H, Komamine A 1986 Effects of nutrient limitation and gamma irradiation on tracheary element differentiation and cell division in single mesophyll cells of Zinnia elegans. Plant Cell Physiol 27:601–606

Sugiyama M, Komamine A 1987 Effect of inhibitors of ADP-ribosyl-transferase on the differentiation of tracheary elements from isolated cells of Zinnia elegans. Plant Cell Physiol 23:541–544

Sung Z R, Fienberg A, Chorneau R, Borkind C, Guiliano G, Nuti Ronchi V 1984 Developmental biology of embryogenesis from carrot cultures. Plant Mol Biol Rep 2:3–15

Sussex I M, Clutter M E 1967 Differentiation in tissues, free cells, and reaggregated plant cells. In Vitro 3:3–12

Suttle J C 1981 Effect of polyamines on ethylene production. Phytochemistry 20:1477–1488

Swain T 1977 Secondary compounds as protective agents. Ann Rev Plant Physiol 28:479–501

Swift H 1950 The constancy of deoxyribose nucleic acid in plant nuclei. Proc Natl Acad Sci USA 36:643–654

Syōno K, Furuya T 1971 Effects of temperature on the cytokinin requirement of tobacco calluses. Plant Cell Physiol 12:61–71

Szweykowska A, Gwóźdź E, Spychala M 1981 The cytokinin control of protein synthesis in plants. In: Guern J, Péaud-Leonoël C (eds) Metabolism and molecular activities of cytokinins. Springer, Berlin Heidelberg New York, pp 212–217

Takahashi N (ed) 1986 Chemistry of plant hormones. CRC, Boca Raton, 282 pp

Talboys P W 1968 Water deficits in vascular disease. In: Kowlowski T (ed) Water deficits and plant growth, vol 2. Academic Press, New York London, pp 255–311

Telewski F W, Jaffe M J 1981 Thigmomorphogenesis: changes in the morphology and chemical composition induced by mechanical perturbation in 6-month-old Pinus taeda seedlings. Can J For Res 11:380–387

Telewski F W, Jaffe M J 1986a Thigmomorphogenesis: field and laboratory studies of Abies fraseri in response to wind or mechanical perturbation. Physiol Plant 66:211–218

Telewski F W, Jaffe M J 1986b Thigmomorphogenesis: anatomical, morphological and mechanical analysis of genetically different sibs of Pinus taeda in response to mechanical perturbation. Physiol Plant 66:219–226

Telewski F W, Jaffe M J 1986c Thigmomorphogenesis: the role of ethylene in the response of Pinus taeda and Abies fraseri to mechanical perturbation. Physiol Plant 66:227–233

Telewski F W, Wakefield A H, Jaffe M J 1983 Computer-assisted image analysis of tissues of ethrel-treated Pinus taeda seedlings. Plant Physiol 72:177–181

Tena M, Lopezvalbuene R, Jorrin J 1984 Induction of phenylalanine ammonia-lyase in hypocotyls of sunflower seedlings by light, excision and sucrose. Physiol Plant 60:159–165

Tepfer D A, Fosket D E 1978 Hormone mediated translational control of protein synthesis in cultured cells of Glycine max. Dev Biol 62:468–497

Theologis A 1986 Rapid gene regulation by auxin. Ann Rev Plant Physiol 37:407–438

Theologis A, Ray P M 1982 Changes in messenger RNAs under the influence of auxin. In: Wareing P F (ed) Plant growth substances 1982. Academic Press, London New York, pp 43–57

Theologis A, Huynh T V, Davis R W 1985 Rapid induction of specific mRNAs by auxin in pea epicotyl tissue. J Molecular Biol 183:53–68

Thimann K V 1937 On the nature of inhibitions caused by auxins. Am J Bot 24:407–412

Thompson N P 1967 The time course of sieve tube and xylem cell regeneration and their anatomical orientation in Coleus stems. Am J Bot 54:588–595

Thompson N P, Jacobs W P 1966 Polarity of IAA effect on sieve-tube and xylem regeneration in Coleus and tomato stems. Plant Physiol 41:673–682

Thomson W W, Platt K A, Campbell N 1973 The use of lanthanum to delineate the apoplastic continuum in plants. Cytobios 8:57–62

Timell T E 1986 Compression wood in gymnosperms, 3 vols. Springer, Berlin Heidelberg New York Tokyo, 706 pp, 1338 pp, 2150 pp

Toister Z, Loyter A 1973 The mechanism of cell fusion II. Formation of chicken erthrocyte polykaryons. J Biol Chem 248:422–432

Torrey J G 1953 The effect of certain metabolic inhibitors on vascular tissue differentiation in isolated pea roots. Am J Bot 40:525–533

Torrey J G 1968 Hormonal control of cytodifferentiation in agar and cell suspension cultures. In: Wightman F, Setterfield G (eds) Biochemistry and physiology of plant growth substances. Runge Press, Ottawa, pp 843–855

Torrey J G 1975 Tracheary element formation from single isolated cells in culture. Physiol Plant 35:158–165

Torrey J G, Fosket D E 1970 Cell division in relation to cytodifferentiation in cultured pea root segments. Am J Bot 57:1072–1080

Torrey J G, Fosket D E, Hepler P K 1971 Xylem formation: a paradigm of cytodifferentiation in higher plants. Am Sci 59:338–352

Tourte Y, Kuligowski-Andres J, Barbier-Raymond C 1980 Comportement différentiel des chromatines paternelles et maternelles au cours de l'embryogenèse d'une fougière: Le Marsilea. Eur J Cell Biol 21:28–36

Tran Thanh Van K, Toubart P, Cousson A, Darvill A G, Gollin D J, Chelf P, Albersheim P 1985 Manipulation of the morphogenetic pathways of tobacco explants by oligosaccharins. Nature 314:615–617

Treshow M ed 1984 Air pollution and plant life. John Wiley, New York, 486 pp

Trewavas A J 1982 Growth substance sensitivity: the limiting factor in plant development. Physiol Plant 55:60–72

Trewavas A J 1983 Nitrate as a plant hormone. In: Jackson M B (ed) Interactions between nitrogen and growth regulators in the control of plant development. Br PGR Group, ARC Letcombe Lab Wantage, pp 97–110

Trewavas A J 1985 Growth substances, calcium and the regulation of cell division. In: Bryant J A, Francis D (eds) The cell division cycle in plants. Cambridge University Press, Cambridge, pp 133–156

Tschermak-Woess E 1956 Karyologische Pflanzenanatomie. Protoplasma 46:798–834

Tucker W Q J, Warren Wilson J, Gresshoff P M 1986 Determination of tracheary element differentiation in lettuce pith explants. Ann Bot 57:675–679

Tukey H B, Went F W, Muir R M, Van Overbeek J 1954 Nomenclature of chemical plant regulators. Plant Physiol 29:307–308

Turing A M 1952 The chemical basis of morphogenesis. Philos Trans R Soc London Ser B 237:37–72

Vallade J, Alabouvette J 1985 Croissance des coenocytes provasculaires dans la racine de Marsilea diffusa (Marsiléales). Can J Bot 63:301–308

van Bragt J, Pierik R L M 1971 The effect of autoclaving on the gibberellin activity of aqueous solutions containing gibberellin A_3. In: van Bragt J, Mossell D A A, Pierik R L M, Veldstra H (eds) Effects of sterilization on components in nutrient media. Veenman and Zonen, Wageningen, pp 133–137

Vanderhoef L N, Dute R R 1981 Auxin-regulated wall loosening and sustained growth in elongation. Plant Physiol 67:146–149

Vanderhoef L N, Kosuge T 1984 The molecular biology of plant hormone action: research directions for the future. Workshop summaries 11. Am Soc Plant Physiol, Rockville, 40 pp

van Overbeek J 1959 Auxins. Bot Rev 25:269–350

van Parijs R, Vandendriessche L 1966 Changes of the DNA content of nuclei during the process of cell elongation in plants II. Variations other than doublings, of the amount of Feulgen stain in maturing plant cells. Arch Int Physiol Biochem 74:581–589

Van't Hof J 1987 Extrachromosomal DNA and cell differentiation in cultured pea roots (Pisum sativum). In: Green C E, Somers D A, Hackett W P, Biesboer D D (eds) Plant tissue and cell culture. Alan R Liss, New York, pp 33–44

Van't Hof J, Bjerknes C A, Lamm S S 1986 Meristematic precursors of vascular parenchyma differentiate from G_2 phase after replicating DNA discontinuously. Am J Bot 73:87–95

Venis M 1985 Hormone binding sites in plants. Longman, London, 191 pp

Verma D P S, van Huystee R B 1970 Cellular differentiation and peroxidase isozymes in cell culture of peanut cotyledons. Can J Bot 48:429–431

Volpi M, Sha'afi R I, Epstein P M, Andrenyak D M, Feinstein M B 1981 Local anesthetics mepracine and propanolol are antagonists of calmodulin. Proc Natl Acad Sci USA 78:795–799

Vonderhaar B K, Topper Y J 1973 Critical cell proliferation as a prerequisite for differentiation of mammary epithelial cells. Enzyme 15:340–350

Vonderhaar B K, Topper Y J 1974 Role of the cell cycle in hormonedependent differentiation. J Cell Biol 63:707–712
von Kaan Albest 1934 Anatomische und physiologische Untersuchungen über die Entstehung von Siebröhrenverbindungen. Z Bot 27:1–94
von Sachs J 1882 Textbook of botany. Morphological and physiological. 2nd edn. Clarendon, Oxford, 980 pp
von Wacek A, Härtel O, Meralla S 1954 Über die Wirkung von Coniferinzusatz auf die Verholzung von Fichtengewebe bei Kultur in vitro. Holzforschung 8:65–67
Waisel Y, Noah I, Fahn A 1966 Cambial activity in Eucalyptus camaldulensis Dehn II. The production of phloem and xylem elements. New Phytol 65:319–324
Walker D B, Bruck D K 1985 The control of positional cell differentiation in plants. In: Sussex I, Ellingboe A, Crouch M, Malmberg R (eds) Plant cell/cell interactions. Current communications in molecular biology. Cold Spring Harbor Lab, New York, pp 53–56
Walton D C 1980 Biochemistry and physiology of abscisic acid. Ann Rev Plant Physiol 31:453–489
Wangermann E 1974 The pathway of transport of applied indoleacetic acid through internode segments. New Phytol 73:623–636
Wardlaw C W 1952 Phylogeny and morphogenesis. Macmillan, London, 536 pp
Wardrop A B 1976 Lignification of the plant cell wall. In: Timell T E (ed) Applied polymer symposia no 28. Proc 8th cellulose conference III. General papers. John Wiley, New York, pp 1041–1063
Wardrop A B 1981 Lignification and xylogenesis. In: Barnett J R (ed) Xylem cell development. Castle House. Tunbridge Wells, pp 115–152
Wareing P F 1982 Determination and related aspects of plant development. In: Smith H, Grierson D (eds) The molecular biology of plant development. Blackwell, Oxford, pp 517–606
Wareing P F, Haney C E A, Digby J 1964 The role of endogenous hormones in cambial activity and xylem differentiation. In: Zimmermann M H (ed) The formation of wood in forest trees. Academic Press, New York, London, pp 323–344
Warren Wilson J 1978 The position of regenerating cambia: Auxin/sucrose ratio and the gradient induction hypothesis. Proc R Soc London Ser B 230:153–176
Warren Wilson J, Dircks S J, Grange R I 1983 Regeneration of sclerenchyma in wounded dicotyledon stems. Ann Bot 52:295–303
Warren Wilson J, Grange R I 1984 Regeneration of tissues in wounded stems: a quantitative study. Ann Bot 53:515–525
Warren Wilson J, Roberts L W, Gresshoff P M, Dircks S J 1982 Tracheary element differentiation induced in isolated cylinders of lettuce pith: A bipolar gradient technique. Ann Bot 50:605–614
Warren Wilson J, Warren Wilson P M 1984 Control of tissue patterns in normal development and in regeneration. In: Barlow P M, Carr D J (eds) Positional controls in plant development. Cambridge University Press, Cambridge, pp 225–280
Watson B, Halperin W 1981 Reinvestigation of the effects of hormones and sugars on xylogenesis in cultured Jersualem artichoke (Helianthus tuberosus) tuber slices, with particular emphasis on the effects of different methods of media preparation and tissue analysis. Z Pflanzenphysiol 101:145–158
Wayne R, Hepler P K 1985 Red light stimulates an increase in intracellular calcium in the spores of Onoclea sensibilis. Plant Physiol 77:8–11
Webber JE, Laver M L, Zaerr J B, Lavender D P 1979 Seasonal variation of abscisic acid in the dormant shoots of Douglas-fir. Can J Bot 57:534–538
Weber H C, Hildenbrand M 1978 Über die sogenannten Phloeotracheiden in den Kontaktorganen von Cansjera rheedii Gmel. (Opiliaceae) und einigen anderen parasitischen Angiospermen. Ber Deutsch Bot Ges 91:231–242
Weiss B, Prozialeck W, Cimino M, Barnette M S, Wallace T L 1980 Pharmacological regulation of calmodulins. Ann NY Acad Sci 356:319–345
Westermark U 1982 Calcium promoted phenolic coupling by superoxide radical – a possible lignification reaction in wood. Wood Sci Technol 16:71–78
Wetmore R H, Rier J P 1963 Experimental induction of vascular tissues in callus of angiosperms. Am J Bot 50:418–430

Weyers J D B, Paterson N W, Brook R A 1987 Towards a quantitative definition of plant hormone sensitivity: opinion. Plant Cell Environ 10:1–10

Whalen M A 1987 Wood anatomy of the American Frankenias (Frankeniaceae): systematic and evolutionary implications. Am J Bot 74:1211–1223

Wheeler R M, Salisbury F B 1981 Gravitropism in higher plant shoots. I. A role for ethylene. Plant Physiol 67:686–690

Wiebe H H, Greer R L, Van Alfen N K 1984 Frequency and grouping of vessel endings in alfalfa (Medicago sativa) shoots. New Phytol 97:583–590

Wilbur F H, Riopel J L 1971a The role of cell interaction in the growth and differentiation of Pelargonium hororum cells in vitro. I. Cell interaction and growth. Bot Gaz 132:183–193

Wilbur F H, Riopel J L 1971b The role of cell interaction in the growth and differentiation of Pelargonium hororum cells in vitro. II. Cell interaction and differentiation. Bot Gaz 132:193–202

Williamson R E 1981 Free Ca^{2+} concentration in the cytoplasm: A regulator of plant cell function. What's New Plant Physiol 12:45–48

Williamson R E 1986 Organelle movements along actin filaments and microtubules. Plant Physiol 82:631–634

Wilson B F 1963 Increase in cell wall surface area during enlargement of cambial derivatives in Abies concolor. Am J Bot 50:95–102

Wilson B F 1986 Apical control of compression wood action in white pine branches. Wood Sci Technol 20:111–117

Włoch W 1981 Nonparallelism of cambial cells in neighbouring rows. Acta Soc Bot Pol 50:625–636

Wodzicki T J, Abe H, Wodzicki A B, Pharis R P, Cohen J D 1987 Investigations on the nature of the auxin-wave in the cambial region of pine stems. Validation of IAA as the auxin component by the Avena coleoptile curvature bioassay and by gas chromatography-mass spectrometry-selected ion monitoring. Plant Physiol 84:135–143

Wodzicki T J, Knegt E, Wodzicki A B, Bruinsma J 1984 Is indolyl-3-acetic acid involved in the wave-like pattern of auxin efflux from Pinus sylvestris stem segments? Physiol Plant 61:209–213

Wodzicki T J, Rakowski K, Starck Z, Porandowski J, Zajączkowski S 1982 Apical control of xylem formation in the pine stem I. Auxin effects and distribution of assimilates. Acta Soc Bot Pol 51:187–201

Wodzicki T J, Wodzicki A B 1980 Seasonal abscisic acid accumulation in the stem cambial region of Pinus silvestris, and its contribution to the hypothesis of a latewood control system in conifers. Physiol Plant 48:443–447

Wodzicki T J, Wodzicki A B, Zajączkowski S 1979 Hormonal modulation of the oscillatory system involved in polar transport of auxin. Physiol Plant 46:97–100

Wolpert L 1970 Positional information and pattern formation. In: Waddington C H (ed) Towards a theoretical biology, vol 3. Edinburgh University Press, Edinburgh, pp 198–230

Wolpert L 1971 Positional information and pattern formation. Curr Top Dev Biol 6:183–224

Wolpert L 1981 Positional information and pattern formation. Philos Trans R Soc London Ser B 295:441–450

Wright K, Northcote D H 1973 Differences in ploidy and degree of intercellular contact in differentiating and non-differentiating sycamore calluses. J Cell Sci 12:37–53

Yamada Y, Sato F, Watanabe K 1982 Photosynthetic carbon metabolism in cultured photoautotropic cells. In: Fujiwara A (ed) Plant tissue culture 1982. Jap Assoc Plant Tissue Cult, Tokyo, pp 249–250

Yamaguichi K, Itoh T, Shimaji K 1980 Compression wood induced by 1-N-naphthylphthalamic acid (NPS), an IAA transport inhibitor. Wood Sci Technol 14:181–185

Yamaguchi K, Shimaji K, Itoh T 1983 Simultaneous inhibition and induction of compression wood formation by morphactin in artifically inclined stems of Japanese larch (Larix leptolepis Gordon). Wood Sci Technol 17:81–89

Yamamoto F, Angeles G, Kozlowski T T 1987a Effect of ethrel on stem anatomy of Ulmus americana seedlings. IAWA Bull N S 8:3–9

Yamamoto F, Kozlowski T T 1987a Effects of flooding, tilting of stems, and ethrel application on growth, stem anatomy and ethylene production of Pinus densiflora seedlings. J Exp Bot 38:293–310

Yamamoto F, Kozlowski T T 1987b Effect of ethrel on growth and stem anatomy of Pinus halepenis seedlings. IAWA Bull N S 8:11–19

Yamamoto F, Kozlowski T T 1987c Effects of flooding of soil on growth, stem anatomy, and ethylene production of Thuji orientalis seedlings. IAWA Bull N S 8:21–29

Yamamoto F, Kozlowski T T, Wolter K E 1987b Effects of flooding on growth, stem anatomy, and ethylene production of Pinus halepenis seedlings. Can J For Res 17:69–79

Yamamoto F, Nakayama T, Suzaki T 1983 Vascular differentiation in callus of Cryptomeria japonica D. Don in vitro. IAWA Bull N S 4:32–38

Yamane H, Takagi H, Abe H, Yokota T, Takahashi N 1981 Identification of jasmonic acid in three species of higher plants and its biological activities. Plant Cell Physiol 22:689–697

Yang S F, Hoffman N E 1984 Ethylene biosynthesis and its regulation in higher plants. Ann Rev Plant Physiol 35:155–189

Yeoman M M 1981 The mitotic cycle in higher plants. In: John P C L (ed) The cell cycle. Cambridge University Press, Cambridge, pp 161–184

Yoshii H, Watanabe A, Imaseki H 1980 Biosynthesis of auxin-induced ethylene in mung bean hypocotyls. Plant Cell Physiol 21:279–291

Yoshimura K, Hayashi S, Itoh T, Shimaji K 1981 Studies on the improvement of the pinning method for marking xylem growth I. Minute examination of pin marks in loblolly pine and other species. Wood Res 67:1–16

Yoshizawa N, Matsumoto S, Idei T 1985 Morphological features of tracheid tips associated with compression wood formation in Larix leptolepis Gord. IAWA Bull N S 6:245–253

Yoshizawa N, Tanaka Y, Idei T 1986 Development of vascular cambium and compression wood formation in the shoot of young spruce (Picea jezoensis var. Hondoensis). IAWA Bull N S 7:21–30

Yu Y B, Yang S F 1979 Auxin-induced ethylene production and its inhibition by aminoethoxyvinylglycine and cobalt ion. Plant Physiol 64:1074–1077

Zajączkowski S, Wodzicki T J 1978 On the question of stem polarity with respect to auxin transport. Physiol Plant 44:122–126

Zajączkowski S, Wodzicki T J, Romberger J A 1984 Auxin waves and plant morphogenesis. In: Scott T K (ed) Hormonal regulation of development II. The functions of hormones from the level of the cell to the whole plant. Encycl Plant Physiol, New Ser, vol 10. Springer, Berlin Heidelberg New York, pp 244–262

Zakrzewski J 1983 Hormonal control of cambial activity and vessel differentiation in Quercus robur. Physiol Plant 57:537–542

Zeroni M, Hall M A 1980 Molecular effects of hormone treatment on tissue. In: MacMillan J (ed) Hormonal regulation of development. I. Molecular aspects of plant hormones. Encycl Plant Physiol, New Ser, vol 9. Springer, Berlin Heidelberg New York, pp 511–586

Zimmermann M H 1978 Hydraulic architecture of some diffuse-porous trees. Can J Bot 56:2286–2295

Zimmermann M H 1982 Functional xylem anatomy of angiosperm trees. In: Baas P (ed) New perspectives in wood anatomy. Nijhoff/Junk, The Hague, pp 59–70

Zimmermann M H 1983 Xylem structure and the ascent of sap. In: Timell T E (ed) Springer series in wood science. Springer, Berlin Heidelberg New York, 143 pp

Zimmermann M H, Brown C L 1971 Trees. Structure and function. Springer, Berlin Heidelberg New York, 336 pp

Zimmermann M H, Potter D 1982 Vessel-length distribution in branches, stems and roots of Acer rubrum L. IAWA Bull N S 3:103–109

Zimmermann M H, Sperry J S 1983 Anatomy of the palm Rhapis excelsa IX. Xylem structure of the leaf insertion. J Arnold Arbor 64:599–609

Zobel R W 1973 Some physiological characteristics of the ethylene-requiring tomato mutant diageotropica. Plant Physiol 52:385–389

Zobel R W 1987 Gaesous compounds of soybean tissue cultures. Carbon dioxide and ethylene evolution. Environ Exp Bot 27:223–226

Zobel R W, Roberts L W 1978 Effects of low concentrations of ethylene on cell division and cytodifferentiation in lettuce pith explants. Can J Bot 56:987–990

Zurfluh L L, Guilfoyle T J 1982 Auxin induced changes in the population of translatable messenger RNA in elongation sections of soybean hypocotyl. Plant Physiol 69:332–337

Appendix: Abbreviations

ABA	abscisic acid
ACC	1-aminocyclopropane-1-carboxylic acid
AMP, ADP, ATP	adenosine phosphates (mono-, di-, and tri-)
AOPP	1-α-aminooxy-β-phenylpropionic acid; inhibitor of PAL activity and lignin biosynthesis
APM	amiprophos-methyl; microtubule depolymerizing agent
ATPase (H^+-ATPase)	ATP-hydrolyzing enzyme; proton (H^+) transporting enzyme system involving ATP energy
AVG	aminoethoxyvinylglycine; ethylene inhibitor
BAP	benzylaminopurine; synthetic cytokinin
CEPA	2-chloroethylphosphonic acid; ethrel; ethephon; ethylene releasing compound
2,4-D	2,4-dichlorophenoxyacetic acid; synthetic auxin; herbicide
DCA	dihydroconiferyl alcohol
DNA	deoxyribonucleic acid
EC	enzyme classification number
G_0	period of cell in a nonproliferating condition; resting state
G_1	phase of the cell cycle following mitosis and characterized by a high rate of RNA formation, protein synthesis, and growth
G_2	phase of cell cycle following DNA replication and prior to mitosis; characterized by 4C level of DNA
GA	gibberellin; the specific compound is indicated by the appropriate subscript
GA_3	gibberellic acid
GC-MS	gas chromatography-mass spectrometry
GC-SIM	gas chromatography-selected ion monitoring
GDP	guanosine 5′-diphosphate
IAA	indole-3-acetic acid; naturally-occurring auxin
2iP	6(γ,γ-dimethylallylamino) purine
2iPA	6(γ,γ-dimethylallylamino) adenosine (2iP riboside)
M	mitotic phase of the cell cycle
M index	mesomorphy index; derived by multiplying the V index by the mean vessel element length
MDMP	D-2-(4-methyl-2,6-dinitroanilino)-*N*-methyl propionamide
MGBG	methylglyoxal *bis*-guanylhydrazone; an inhibitor of S-adenosyl-methionine decarboxylase
MT	microtubule(s)
NAA	α-naphthaleneacetic acid

NAD	nicotinamide adenine dinucleotide
NADP	phosphorylated NAD
P_{730}	far-red absorbing form of phytochrome pigment
PAL	L-phenylalanine ammonia-lyase
PEG	polyethylene glycol
PP_2 (P-protein)	phloem-localized protein
QC	quiescent center
RNA (rRNA, mRNA)	ribonucleic acid; the r prefix indicates ribosomal RNA; the m prefix indicates messenger RNA
S, S phase	period of the cell cycle during DNA replication
S_1, S_2, S_3	secondary wall layers deposited successively on the inner surface of the primary wall during secondary wall thickening
SAM	S-adenosylmethionine; intermediate in the biosynthesis of ethylene
TDF	tracheid differentiation factor
TE	tracheary element(s)
TIBA	2,3,5-triiodobenzoic acid; inhibitor of polar auxin transport
UDP	uridine diphosphate
V index	vulnerability index; calculated by dividing the mean vessel diameter by the number of vessels per mm^2
w/v	percent "weight in volume"; number of grams of constituent in 100 cm^3 of solution
w/w	percent "weight in weight"; number of grams of constituent in 100 g of matrix

Subject Index